A Simplified Guide to Statistics
for Psychology and Education

Fourth Edition

G. Milton Smith
The City College of the City University of New York

HOLT, RINEHART AND WINSTON, INC.
New York Chicago San Francisco Atlanta
Dallas Montreal Toronto London Sydney

To *Frances, Winthrop, David, Anne, and Héloise,*
who, on the basis of random sampling,
are highly improbable.

Copyright 1938, 1946 by G. Milton Smith
Copyright © 1962, 1970 by Holt, Rinehart and Winston, Inc.
All rights reserved
Library of Congress Catalog Card Number: 73-95568
SBN: 03-082815-5
Printed in the United States of America
1 2 3 4 5 6 7 8 9

Preface

Many statistical concepts and techniques are today universally recognized as essential equipment for students of the behavioral sciences. This is especially true in the fields of psychology, education, and sociology. The principal purpose of this *Guide* is to bring together and integrate the more commonly used statistical tools and concepts and to make them meaningful. Earlier editions have proved useful in general and laboratory courses and as research aids.

This enlarged fourth edition may be used also as an easy text in short introductory statistics courses for psychology, education, and other behavioral sciences. Its style and the use of simplified concrete illustrations should make it helpful as an explanatory supplement in statistics courses using more difficult texts.

The new edition retains most of the features of the third edition, but with some modifications. The chapters on the normal curve and significance tests for both large and small samples have been completely reorganized. New figures and tables have been introduced to clarify probability concepts and to explain the use of one-tailed and two-tailed tests. To the chapter on correlation and regression equations a section on the phi coefficient of correlation has been added. A formula for correction for discontinuity has been added to the chi-square chapter. The formulas for the variance and standard deviation of samples are now given with $n - 1$ in the denominators (rather than with n) to correct for bias. These formulas, for both original and coded scores, have been introduced with emphasis on the sum of squares, Σx^2, which makes possible a meaningful tie-up with the new chapters on analysis of variance.

Five wholly new chapters deal with simple analysis of variance (with sections on Scheffé comparisons), two-factor variance analysis, the power of significance tests, decision making, including Type I and Type II errors, and two easy "nonparametric" significance tests, the median test and the Kruskal-Wallis rank test. The new subjects, sometimes baffling to the beginner, have been made generally comprehensible and occasionally entertaining. This edition also includes new, compact, and easily read tables of z, t, and F, and answers to exercises.

In earlier editions I have already acknowledged my indebtedness to Professor H. E. Garrett's lucid teaching and to his *Statistics in Psychology and Education*, and my debt for many helpful suggestions to Professor H. H. Abelson, formerly Dean of the School of Education of the City University of New York, Professor Charles Bird, formerly of the University of Minnesota, Professor E. E. Cureton of the University of Tennessee, Professor Allen L. Edwards of the University of Washington, and Dr. Norman Frederiksen, for many years Director of Research at the Educational Testing Service in Princeton. In the work of revision for the last three editions I have found most valuable the late Sir Ronald A. Fisher's *Statistical Methods for Research Workers*, William L. Hays's *Statistics for Psychologists*, E. F. Lindquist's *Statistical Analysis in Educational Research*, Croxton and Cowden's *Applied General Statistics*, Peters and Van Voorhis's *Statistical Procedures and Their Mathematical Bases*, H. M. Walker's *Elementary Statistical Methods*, P. G. Hoel's *Elementary Statistics*, and Allen L. Edwards's *Statistical Analysis* and *Statistical Methods* (second edition). Dr. Eleanor W. Willemsen of the University of California at Berkeley and Dr. David Louis Brown of Columbia University have gone over the manuscript of this new edition in detail. I have gratefully adopted many of their suggestions.

G. M. S.

New York, N. Y.
September 1969

Contents

PREFACE — iii

1 EXPERIMENTATION AND TESTING; THE NEED FOR STATISTICS — 1
 1. The Statistical Nature of Science — 1
 2. Experimentation in Physics and Psychology Compared — 2
 3. The Control Group — 4
 4. The Desirability of Using a Fairly Large Number of Subjects in Psychological and Educational Experimentation and Testing — 5
 5. Experimentation and Testing Contrasted — 6

2 DISTRIBUTION OF SCORES — 8
 1. Scores for a Continuous Variable — 8
 2. Tally and Frequency Table — 9
 3. Graphical Representation of a Frequency Distribution — 12
 4. A Normal Probability Curve (or Normal Distribution Curve) — 14
 Exercises — 17

3 MEASURES OF CENTRAL TENDENCY — 19
 1. The Mean — 20
 2. The Crude Mode — 22
 3. The Median — 22
 4. The Median Is Sometimes More Representative than the Mean — 24
 5. Short Methods for Calculating the Mean by Coding — 26
 *6. Derivation of the Two Formulas for Calculating the Mean by Coding; Three Rules of Summation — 30
 Exercises — 32

4 MEASURES OF VARIABILITY — 33
 1. The Range — 35

	2. The Mean Deviation (MD)	35
	3. The Standard Deviation (s) and the Variance (s^2)	36
	4. Methods for Obtaining the Standard Deviation (s) and the Variance (s^2) by Calculating Machine and by Coding	39
	5. Accuracy of Calculations: Charlier's Check	44
	6. The Quartile Deviation (Q)	44
	7. The Interrelation of Q, MD, and s	46
	*8. Derivation of Special Formulas for the Sum of Squares Required for the Variance and Standard Deviation	47
	Exercises	49
5	**THE USE OF NORMS AND GRADING "ON THE CURVE"**	**51**
	1. The Significance of Scores Is Relative	51
	2. Quartile, Decile, and Percentile Norms	52
	3. Grading "On the Curve"	53
	Exercises	55
6	**STANDARD SCORES FOR COMPARING AND COMBINING TEST RESULTS**	**56**
	1. Standard Scores (z Scores)	57
	2. T Scores	59
	3. Percentile Rank and Order of Merit Equivalents	61
	Exercises	62
7	**PROBABILITIES DETERMINED FROM A NORMAL DISTRIBUTION, THE z DISTRIBUTION; CONFIDENCE LIMITS FOR THE MEAN**	**64**
	1. Areas under the z Distribution	65
	2. The Variance and Standard Error of the Mean; Confidence Limits for the Mean	70
	Exercises	73
8	**THE SIGNIFICANCE OF A DIFFERENCE BETWEEN THE MEANS OF SMALL SAMPLES, THE t TEST**	**75**
	1. The Significance of a Difference between the Means of Two Small Independent Samples; the t Test	76
	2. The Calculation of t for Small Independent Samples of Unequal Size	82
	3. The Significance of a Difference between the Means of Two Correlated Samples	83
	Exercises	87
9	**A SIGNIFICANCE TEST FOR LARGE SAMPLES; ONE-TAILED AND TWO-TAILED TESTS**	**89**
	1. The Significance of a Difference between the Means of Two Large Independent Samples	89

Contents vii

    2. One-Tailed and Two-Tailed Significance Tests............ 90
    Exercises.. 95

*10 DECISION MAKING AND THE POWER OF SIGNIFICANCE TESTS   **98**
    1. Decision Making and Calculating Test Power............ 99
    2. How To Increase the Power of a Significance Test........ 106
    Exercises.. 111

11 SIMPLE ANALYSIS OF VARIANCE............................. **113**
    1. Main Steps in Simple Analysis of Variance.............. 114
    2. The General Nature of the F Ratio and the F Table...... 115
    3. Interpretation of the F Ratio......................... 120
    4. Calculation of the Three Sums of Squares for Simple Analysis of Variance.................................... 121
    5. Summary of Simple Analysis of Variance; Calculation and Significance of F... 124
    6. The Scheffé Test of Significance for a Difference between Any Two Means... 125
    *7. The Scheffé Test for More Complex Comparisons......... 128
    Exercises.. 131

12 TWO-FACTOR ANALYSIS OF VARIANCE........................ **133**
    1. Stage I. Preliminary Analysis........................... 135
    2. Stage II. Further Analysis of the Variance between Means.. 138
    3. A Two-Factor Three-Level Experiment on Drug Effects... 145
    *4. Fixed Effects and Other Experimental Designs........... 151
    Exercises.. 152

13 CORRELATION TECHNIQUES; RELIABILITY OF TESTS AND CONFIDENCE LIMITS FOR TEST SCORES........................... **154**
    1. The Rank-Difference Correlation Method................ 155
    2. The Product-Moment Method of Correlation............. 157
    3. Interpretation of Correlation Coefficients................ 159
    4. Reliability and Validity of Tests......................... 160
    *5. Confidence Limits for Individual Test Scores............. 162
    Exercises.. 163

14 MACHINE AND CHART CORRELATION METHODS; PREDICTION FROM REGRESSION EQUATIONS; THE PHI COEFFICIENT OF CORRELATION.. **165**
    1. Machine Calculation of r from Ungrouped Original Scores... 165
    *2. Calculation of r from Coded Grouped Data by Means of a Correlation Chart...................................... 167
    3. Predictions from Correlations by Means of Regression Equations and the Standard Error of Estimate........... 173

Contents

*4. Predictions from Machine and Chart Correlations Compared ... 177
5. Reduction in Error of Estimate as a Function of r 179
6. The Phi Coefficient of Correlation 180
Exercises 183

15 THE χ^2 DISTRIBUTION FOR TESTING HYPOTHESES 185

1. Calculation of the χ^2 Statistic; Goodness of Fit 186
2. Interpretation of χ^2; Use of χ^2 Table 187
3. Degrees of Freedom in χ^2 Tests 189
4. The Use of χ^2 in Tests of Independence (and Association) ... 191
5. Correction for Discontinuity 197
Exercises 198

16 SIGNIFICANCE TESTS BASED ON THE ORDER OF SCORES 201

1. The Kruskal-Wallis Rank Test of Significance 202
2. The H Test Compared with the F and t Significance Tests ... 206
3. The Median Test 207
4. The Median Test Compared with Other Significance Tests 212
Exercises 213

SELECTED ANSWERS TO EXERCISES 233

INDEX 239

List of Tables

1. Tally and Frequency Table for 42 Intelligence Test Scores...... 10
2. Calculation of the Mean, the Crude Mode, the Median, and the Quartile Deviation from Grouped Data...... 20
3. Calculation of the Mean of Large Ungrouped Scores Coded by Subtraction...... 27
4. Calculation of the Mean of Grouped Scores Coded by Subtraction and Division...... 29
5. Calculation of the Mean Deviation (MD), the Standard Deviation (s), and the Variance (s^2) Illustrated for Ungrouped Data...... 37
6. Calculation of the Variance (s^2) and the Standard Deviation (s) from Original Scores Ungrouped...... 41
7. Calculation of the Variance (s^2) and the Standard Deviation (s) from Coded Grouped Data...... 43
8. Converting Test Scores into Equivalent T Scores...... 60
9. Combining Test Scores by Averaging Order of Merit Rankings.... 62
10. The Distribution of Areas under the Normal Probability Curve Corresponding to Distances on the Baseline between the Mean and x/σ 68
11. Distribution of t 80
12. A Test of Significance for the Difference between the Means of Two Correlated Samples (a Check on the Apparent Loss in the Perception of Speech at an Altitude of 16,900 Feet)...... 85
13. The $p = .05$ and $p = .01$ Points for Selected Distributions of F...... 116
14. Simplified Data for $k = 3$ Treatment Groups, with $n = 5$ Subjects in Each Group Selected at Random...... 122
15. Simplified Scores, Means, and Sums for a 2×2 Factorial Experiment...... 136

List of Tables

16. Sums for the Four Factor Combinations of Table 15 Arranged in a 2 × 2 Table...138

17. Group Means and Level Means for a 2 × 2 Factorial Experiment Based on the Sums in Table 16..............................141

18. Sums for the Nine Factor Combinations for a 3 × 3 Factorial Experiment on Two Drugs...................................146

19. Treatment Means for a 3 × 3 Factorial Experiment on Two Drugs 149

20. Rank-Difference Correlation Technique Illustrated.............156

21. Product-Moment Correlation Technique Illustrated............158

22. Calculation of the Product-Moment Correlation (r) from Coded Grouped Data — A Correlation Chart........................169

23. Reduction in the Standard Error of Estimate as a Function of r 181

24. Calculation of χ^2 for Testing Goodness of Fit between Observed and Theoretical (Normal) Distributions of 200 Grades in Elementary Physics..186

25. Fisher's Table of χ^2...188

26. Calculation of χ^2 for a 2 × 2 Table in a Test of Independence......192

27. The χ^2 Test of Independence Applied to the Hypothesis that High School Grades Are Independent of Social Adjustment...........194

28. Values of $(f_o - f_t)^2/f_t$ Calculated from the Data of Table 27, Suggesting Why Grades Are Not Independent of Social Adjustment...196

29. The Calculation of H from Ranked Scores for Three Groups.....204

30. The Calculation of H from Ranked Scores for Two Groups of Unequal Size...205

31. A Median Test for Four Treatment Groups Using a χ^2 Test of Independence in a 2 × 4 Table..............................210

32. Table of Squares and Square Roots of Numbers from 1 to 1000..216

1
Experimentation and Testing; the Need for Statistics

1. The Statistical Nature of Science

At its best science is statistical. This means that "scientific laws" do not express *with certainty* how nature behaves: scientific laws merely describe how nature *has* behaved *within limits* and how it is *likely* to behave within limits again, under similar conditions. It is the business of statistics to determine what these limits are, under any particular set of conditions, and to work out the probability of the recurrence of any given set of events, in the light of the frequency and the regularity of their occurrence in the past.

We cannot say with certainty that the sun will rise tomorrow. But we can say that, without recorded exception, it *has* risen each day since the beginning of history, and that the more obvious conditions of its rising have not changed appreciably; it is, therefore, *highly probable* that it will rise tomorrow. Likewise, judging from the past, it is probable that, in northern latitudes, the first of January in any future year will be colder than the first of April. We cannot be certain of this, however; for New Year's Day is sometimes comparatively mild and it has been known to snow on April first. Since there is considerable variation in these matters, it is not beyond the bounds of possibility that a phenomenally mild January first and a phenomenally chilly April first will some time in the future occur in the same year. If this should happen, the normal temperature relations would be reversed. The probability of such a reversal, however slight, is clearly greater than that of the sun's developing a grouch and refusing to rise some morning. We say this with some confidence because of the contrast between the constancy of the sun's movements in the past and the fickleness of temperature conditions. Thus we see that *degree of probability* of future recurrence is directly tied up with degree of constancy, or lack of variability, in past occurrence. Statistics deals with *averages* of past events, *variability* about these averages, and the *probability* of future events conforming to past averages. Statistics is also very much concerned with estimating the characteristics of large groups on the basis of smaller samples and with special techniques for evaluating experimental results. But before we get to these matters, let us see why we need statistical tools in psychology and education even more, perhaps, than in the so-called exact sciences.

2. Experimentation in Physics and Psychology Compared

From the statistical point of view, the essential difference between psychology and such sciences as physics and chemistry is one of variability. Psychological phenomena exhibit vastly greater variability than do physical phenomena, and they are correspondingly less predictable. This is so because of the enormously greater number of variables which enter into the determination of human and animal behavior. A comparison of a familiar experiment in physics and a relatively simple experiment in psychology will bring this out.

In the usual check on Boyle's law in the physics laboratory

we are interested in the relationship of volume to pressure in a confined gas. We want to know how volume changes as the result of systematic variation in pressure. There is only one other important variable with which we need concern ourselves—temperature. It is clear that, if we do not keep temperature constant, changes in volume may result from temperature changes as well as from pressure changes, and we shall not know how much of the result to attribute to each. But it is a relatively simple matter to keep this third variable constant during the experiment. Changes in volume can then be definitely tied up with pressure changes, and a law governing their relationship, within the limits of error of the experiment, can be formulated.

In contrast with this clear-cut three-variable problem from the inorganic field, let us now consider a problem from the psychological laboratory. Suppose we wish to test the effect of alcohol on mental performance. To make it specific, let us say we wish to determine the effect of three ounces of the beverage on performance in an addition test, fifteen minutes after indulgence. In rough outline, the usual procedure would be to give the subjects an initial test; introduce the experimental factor (in this case, the alcohol); and then, after the specified interval, give a final test. Now, if the alcohol has any effect at all, we should expect a difference between the initial and the final test scores. But can we attribute any difference which may occur to the experimental factor alone? Unfortunately not; for, without exhausting the possibilities, we can list a large number of other variables which might be responsible for some part of the change: fatigue, practice, unequal difficulty of the two tests, unequal distractions due to noise or to variations in illumination, warming up to the task, cooling off from boredom, suggestion based upon preconceived notions of what alcohol does, and so on. In the face of this formidable array of extraneous influences, it may seem a hopeless task to isolate the effect of the intended experimental factor, the alcohol. It is not hopeless, however; for allowance may be made for many of these interfering variables by means of a "control group." The use of a control group, an essential device in most psychological experimentation, deserves special mention.

3. The Control Group

The customary procedure is to divide the available subjects into two groups, the experimental group and the control group, *equated* as nearly as possible with respect to age, sex, general intelligence, previous experience with the experimental factor, and with respect to performance on the initial test (or tests). Equating may be carried out in several ways: by *pairing* the subjects in the two groups with respect to all relevant traits; by simply *equating the averages* of the various traits; or by having *the same proportion* of subjects with the various traits in each group. (The groups do not have to be of the same size, though this is sometimes desirable. Pairing would of course insure groups of equal size.) Throughout the experiment the two groups are treated, so far as possible, in just the same manner, with one exception. Each takes the initial test and, after a uniform delay, each takes the final test. The only intentional difference is that the experimental factor (usually concealed in some way in order to eliminate suggestion) is introduced in the experimental group but not in the control group. Then, if the change in scores between initial and final tests is greater for one group than for the other, this is presumably due either to certain chance factors or to the influence of the experimental factor, since the two groups were subject to the same amount of work, practice, distraction, and so on. A *small* control group does *not* make adequate allowance for such chance factors as inequalities in the subjects' alertness, perseverance, and fatigability. Such factors as these are of great importance and may even wholly submerge the effect of the experimental factor if the groups are small. On the other hand, if the groups are large and are selected without bias, the influence of the chance factors tends to become smaller, and the effect of the experimental factor may be more readily determined. This point is amplified in the following section.

It should be emphasized that the use of the control group is not a foolproof experimental procedure. Even when a clear difference appears between the performances of the experimental and the control groups, we cannot be sure that this is due exclusively to the experimental factor. It is always possible that there is some obscure factor, overlooked by the experimenter, which is being introduced unintentionally along with the experimental factor, and which may be even more influential than the experimental factor intentionally introduced. Strictly speaking, the use of a properly selected control group does no more than assure the experimenter that any differences he may find between his experimental and control groups, following the introduction of the experimental factor, are *not* due to the factors which are equal for the two groups.

4. The Desirability of Using a Fairly Large Number of Subjects in Psychological and Educational Experimentation and Testing

Although there are certain areas of experimentation in which it is not necessary to have large numbers of subjects in order to obtain worth-while results (see Chapter 8), very small samples are usually undesirable and may yield meaningless or misleading results. To illustrate this, let us take the extreme case in which an experimental group composed exclusively of John Jones is to be compared in ability to add with a control group made up exclusively of Jim Smith, before and after John takes three ounces of alcohol (to return to our illustration above). A difference in performance may be due more to chance differences in the make-up and experience of the two subjects than to the experimental factor. For example, though they have been selected because they are of the same age and sex, and have obtained approximately equal scores on some intelligence test, John may be considerably heavier than Jim. Hence John would receive less alcohol per unit of body weight than Jim would have received, had he happened to be the experimental group. Jim may be more easily bored than John. John may be an habitual toper and Jim a teetotaler; this would have an important bearing on their relative susceptibility to the drug. Jim may be a grocery clerk, and daily practice in adding may have brought him close to his limit of skill in this function; John may have a profound detestation for arithmetic in any form. Many of these variables are intangible and none of them is covered by the use of the control group. Thus, any conclusion about the effects of alcohol based on such a study would probably be erroneous. We should not know how much of the effect to attribute to the experimental factor and how much to the other factors. Even if we could separate these influences, our conclusions about alcohol would still have to be confined to John Jones, for we do not know that he is typical however frequently he may occur.

If, on the other hand, our experimental and control groups each contain 50 or 100 subjects, or even more, we can assume that, taken as a whole, they are similar with respect to most of the noncontrollable factors suggested above. And we can be fairly confident that, within each group, the positive deviations in most of these matters will cancel out the negative deviations. The larger the groups are, the smaller will be the probable distortion by the chance factors, and the greater the likelihood that the findings will apply generally to similar groups.

It should now be apparent that, if the results of psychological experimentation are to have general applicability, we shall ordinarily have to deal with large numbers of test scores or other measures. This applies to educational experimentation as well. The need, therefore, for tools for organizing and analyzing our data becomes clear. The more important of the tools which statistics provides for these purposes are described in the chapters that follow, along with important techniques for determining the significance of experimental and test findings.

5. Experimentation and Testing Contrasted

In psychological and educational experimentation we often make use of tests of one sort or another, as suggested in the discussion above; but tests are often used also for their own sake, with no experimental purpose in mind. One important point of contrast between experimentation and testing may be roughly expressed as follows: in experimentation we keep our subjects constant and vary the conditions; in testing we keep the conditions constant and vary the subjects. That is, in experimentation we give our initial and final tests to the *same* group (or groups) of subjects; to do otherwise would produce meaningless results. The introduction of the experimental factor constitutes the variation in conditions. In testing, on the other hand, our purpose is to apply a standard measuring stick to a variety of subjects under constant conditions. If the conditions are not kept constant, we shall not know whether a difference between two scores, or sets of scores, is due to a difference in ability between the two individuals or groups compared or to the change in test conditions.

As in experimentation, the results of testing will be more valuable if many subjects are used; and when many scores are to be dealt with, we must call on statistics for tools of analysis. Statisticians have also developed sophisticated techniques for making limited predictions from relatively small samples. (See Chapter 8.)

2
Distribution of Scores

1. Scores for a Continuous Variable

Most of the traits which we are interested in measuring in psychology and education may be thought of as *continuous variables*. A continuous variable is without gaps and is theoretically divisible into an unlimited number of subdivisions. For example, height and weight are commonly measured, not just in units which are discrete integers, but in fractions of inches or of pounds as well. And although test scores are usually expressed for convenience as whole numbers, there is no theoretical reason why they could not be expressed in smaller units. Certainly an average of several scores expressed in whole numbers and fractions thereof is a perfectly commonplace and meaningful quantity.

For a continuous variable *scores* are to be regarded, not as points on a scale, but as *distances along a scale* between two limiting points. In the literature of psychology and education there are, unfortunately, two common methods of defining the limits of the interval represented by a single score. In one, a score of 37, for example, stands for all values between 36.5 and 37.5, as in the insurance companies' definition of "age 37." The midpoint of this score interval would be 37.0. In the other, a score of 37 stands for all values from 37.0 to 38.0, as in the layman's definition of "age 37." The midpoint of this interval would be 37.5. The two definitions differ by 0.5 of a point only. The former definition, which is mathematically more precise and now more commonly used, is the one employed throughout this manual.

2. Tally and Frequency Table

Let us suppose that an instructor gives his students an aptitude test, partly with the idea of illustrating the testing procedure, and partly with the idea of sizing up the abilities of the class. He wants to get a rough idea as to whether he will be teaching just another average class, whether he will have to struggle to get his ideas across, or whether he will have to be on his guard against an excessively bright group of students who will be constantly catching him up on careless slips. He also wants to get a rough notion as to whether the class is of fairly uniform ability or whether there will be a large and annoying difference between the worst and the best students, with the result that what he says may pass over the heads of some but may bore the brighter members of the class.

Here is a sample of the scores he might obtain:

Aptitude Test Scores for 42 College Students
(UNGROUPED DATA)

61	60	50	52	58	38	60	51	55	*68*	55	62	47	39
58	42	47	42	48	49	48	46	55	51	58	65	45	35
43	54	52	56	46	65	53	*34*	48	50	39	59	53	52

Now what do these scores reveal? As they stand—nothing. Some sort of organization is called for. Order may be quickly brought out of chaos by constructing a *frequency table*. This is done by dividing the total range of scores into a convenient number of equal steps, or classes, and then sorting out the scores by placing a mark, or *tally*, for each score opposite the step in which it falls. This procedure is illustrated in Table 1.

Here two common methods of designating the steps or classes are given. The one on the left has the advantage of compactness and may facilitate accuracy of tabulation. It is preferable when dealing with test scores which do not involve fractional values, *provided* the actual limits of the steps are kept in mind. The alternative method gives a clearer indication of the actual limits of the steps. It also gives a clearer indication of the *step interval*, or *class interval*, i.e., the distance along the scale corresponding to a single step or class. In this case it is 3 score

TABLE 1

Tally and Frequency Table for 42 Aptitude Test Scores
(GROUPED DATA)

Step (or Class)		Tally	f
Simpler Designation	Alternative Designation (Actual Limits)		
[69–71	68.5–71.5		0]
66–68	65.5–68.5	/	1
63–65	62.5–65.5	//	2
60–62	59.5–62.5	////	4
57–59	56.5–59.5	////	4
54–56	53.5–56.5	////	5
51–53	50.5–53.5	//// //	7
48–50	47.5–50.5	//// /	6
45–47	44.5–47.5	////	5
42–44	41.5–44.5	///	3
39–41	38.5–41.5	//	2
36–38	35.5–38.5	/	1
33–35	32.5–35.5	//	2
[30–32	29.5–32.5		0]
			$n = 42$

units, and not just 2 as it might appear to be from a hasty glance at any one step indicated by the simpler method. In the tables in subsequent chapters we shall usually use the simpler method of designating the steps or classes. To avoid confusion in determining the step interval, or length of step, with this method the safest procedure is to take it as the difference between the beginning of any one step and the beginning of the next.

A word should be said here about the *choice of step interval* when one starts from scratch with an unorganized set of scores like those above. There is no fixed rule for this, but the following suggestions may be helpful. For purposes of calculation, it is *usually* desirable to divide the total *range* of scores, high score minus low score (in the above data, $68 - 34 = 34$), into a number of steps which is not less than 10 or greater than 20. Certain values for the step interval are more convenient than others, e.g., 3, 5, 7, because as odd numbers they will yield a step midpoint which is a whole number rather than one ending in .5; and 10 or 20 because we so commonly work with multiples of 10. It is common, though not essential, practice to choose as the beginning of the first step in a frequency table a number which is a multiple of the step interval, as in Table 1.* When the scores are to be presented in graphical form (see section 3) it is sometimes desirable to work with a smaller number of steps of somewhat greater length than is used for purposes of calculation.

Let us now return to a consideration of Table 1. The labeling of the steps and the tally have already been discussed. The top and bottom rows in brackets will be explained in the section which follows. The letter f in the table stands for frequency; the column beneath it is called the *frequency column*. It is simply the summary of the tally. Once we have our frequency column filled in, we may discard the tally; what remains

---

*It sometimes happens that scores tend to occur more or less spontaneously *in groups at regular intervals*. For example, in grading essay examinations teachers use most commonly such grades as 65, 70, 75, 80, etc., and much less commonly such grades as 68, 77, 83, etc. In cases of this sort, "*constant grouping errors*" may be minimized by choosing the step interval and the position of the step in such a way as to *center within the steps* the grouped scores. In this case the steps should be 63–67, 68–72, etc.

is the frequency table proper, one form of frequency distribution. The letter *n* stands for total frequency, or total number of cases.

Now that we know how the frequency table is constructed, let us see what this particular frequency table has to tell about the instructor's class. Though we have not yet finished our analysis (and can tell nothing, for example, about the general level of ability, which must be found by comparison with previously determined points of reference, or *norms*, discussed in Chapter 5), we are already in a position to answer one of the instructor's questions. We can see that 23 scores representing more than half the class fall in the four middle steps, within a range of 12 points. On both sides of these middle steps the frequencies drop off gradually and with fair symmetry toward the end steps, where the extreme scores are few in number. It is, after all, a class which is fairly homogeneous with respect to the traits measured by the test. If the instructor gauges his remarks for the students in those four middle steps, he will not be far from the capacities of those on either side. In any case, he does not have to deal with a class which is split into two clear-cut groups, one very much more capable than the other.

The general nature of any frequency distribution can be grasped more quickly if the facts contained in a frequency table are portrayed graphically. Two common methods of graphical representation are described in the next section.

3. Graphical Representation of a Frequency Distribution

A frequency distribution is commonly graphed by means of either a frequency polygon or a histogram. These are illustrated in Figures 1 and 2, respectively, which are based on the data from the frequency table in the section above. In both cases the steps are conventionally indicated on the horizontal axis and the frequencies on the vertical axis. In the construction of the *frequency polygon* the points (representing the number of cases or scores within the respective steps) are plotted above the *midpoints* of the steps and aligned with the appropriate fre-

Graphical Representation of a Frequency Distribution

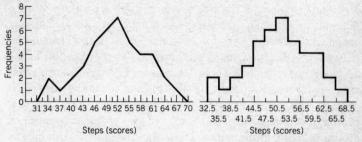

Figure 1. Frequency polygon.

Figure 2. Histogram.

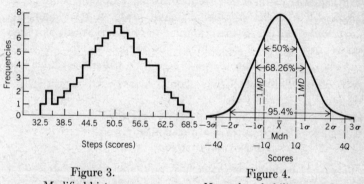

Figure 3. Modified histogram.

Figure 4. Normal probability curve.

quencies. It is customary to plot an *extra point* on the base line at each end of the figure, making use of an extra step at each end of the frequency table, for which the frequency is 0 (see steps in brackets in Table 1). The figure is completed by connecting the adjacent points by *straight* lines. The *histogram* is the "sky line" of the figure formed by erecting a rectangular column on each step in such a way that the breadth of each column is equal to the step interval and its height is equal to the frequency.

Let us examine for a moment these two figures based upon our aptitude test data. They do not really contain any more information than the frequency table; they simply make the facts stand out more clearly—in this case, the piling up of the frequencies in the center and the fairly symmetrical tapering off toward the higher and lower score values (or steps).

The functions of the two types of graph are not essentially different. Except for purposes of illustration, we should not ordinarily construct both for the same set of data. Though the frequency polygon is smoother than the histogram, the histogram has one advantage. It is easier to see a relationship between areas under various parts of the histogram (or under the entire histogram) and frequencies. For example, we can think of each column of the histogram as being made up of the same number of *unit areas* on graph paper (usually squares) as the frequency of the step in question. Then *the total area under the histogram* will equal the sum of the frequencies for all of the columns, n, expressed in terms of the same unit areas. Furthermore, the *area under any particular section* of the histogram can be taken as representative of the sum of the frequencies for the steps involved. Some applications of these facts will appear later (see especially Chapter 7).

4. A Normal Probability Curve (or Normal Distribution Curve)

This section does not make any pretense at completeness of proof of the statistical theory it presents; it is merely intended to make more understandable some of the concepts discussed in subsequent chapters.

In Figure 3 we have suggested the probable form which the histogram of Figure 2 would take if n were doubled. The number of steps has been increased to accommodate the new cases, and the size of the steps has been cut in half. If this process were to be repeated again and again, it is not difficult to see what would happen: the outlines of the histogram would more and more closely approximate the curve in Figure 4, so that, as n approached infinity and the size of the steps approached zero, the two figures would tend to become one. We can thus think of a normal probability curve as an *idealized* graph of certain common frequency distributions* for which n is very large. It will

---

*Less commonly we may, under special circumstances, get so-called **J** or **U** distributions.

not necessarily have just these proportions; it may be considerably flatter or it may be taller and thinner; but it will always be symmetrical and bell shaped.

A curve of this general shape is called a *normal probability curve* because it describes, among other things, the most probable distribution of the frequencies of certain chance events. (It is sometimes called "the curve of chance.") For example, if we toss 8 coins several thousand times, it can be shown that, *on the average*, the various possible combinations of heads and tails will occur with the following frequencies:

Probable Frequency of Various Combinations in Tossing 8 Coins

Heads	Tails	Number of Occurrences[a] in 256 Tosses
8	0	1
7	1	8
6	2	28
5	3	56
4	4	70
3	5	56
2	6	28
1	7	8
0	8	1
		Total = 256

[a]The values in this column are simply the coefficients of the terms which result from the expansion of the binomial $(H + T)^8$; i.e., $1H^8 + 8H^7T + 28H^6T^2 + \ldots + 1T^8$. The number 256 is the sum of these coefficients.

These values may be checked by an actual tossing of the coins. Their symmetry is obvious. If we plot them, we have a fair approximation to a normal probability curve.* If we work with 16 or 32 coins (and with considerably more patience; for with 32 coins we cannot expect them to fall all heads or all

*Actually what we get from this procedure is what is known as a *binomial distribution;* but the two distributions are very similar for 10 or more coins.

tails more than once in approximately every 4,293,000,000 times), the approximation to a normal curve becomes very close indeed.

This curve is of great importance, not only for statistics (which bases many of its techniques of analysis upon it), but for science in general, for it pictures the normal, or most common, distribution of measures for a surprising array of phenomena, psychological and otherwise, when the data are derived from *a large and unselected sample* (i.e., chosen at random or by chance). For example, the frequency distributions for height and weight, average monthly temperature and rainfall over a long period of years, and the like, all tend to approximate this curve. We have just seen how closely our distribution of the aptitude test scores approximated the curve. (This close approximation cannot, however, always be expected, particularly with such a small group of subjects.)

The most common deviation from a normal curve that we meet in practice is known as *skewness*. That is, the curve is not symmetrical, but is skewed to one side or the other, the higher frequencies tending to pile up nearer one end rather than in the middle. Moderate amounts of skewness do not seriously affect the statistical quantities discussed in later chapters, even though many of these are derived on the assumption that the frequency distribution is normal. A deviation from a normal probability curve which would be serious is a *bimodal* distribution; that is, a two-humped curve, like the back of an Asiatic camel. This would mean that the group to be studied was far from homogeneous, that it was really two groups. For example, an intelligence test given to a group made up partly of subjects from an institution for the feeble-minded and partly of students from almost any college would produce such a curve. Clearly any generalization made about the average of such a group would be worthless, for the average would fall between the two humps and would represent practically no one in the group.

Because the distributions of scores for so many measured human traits have approximated a normal probability curve so often in the past, we are a little suspicious when we get a marked deviation from this pattern. We are inclined to suspect that selective factors have entered into the choice of our subjects: such factors as unequal age distribution or unequal educational background. Or perhaps a test is too easy for a group. In such a case the distribution curve would be skewed so as to end too abruptly above the mean. Similarly, if a test is too difficult, scores tend to bunch up near the lower end of the curve. In any case, before we go far with an experiment or a testing program, we should check, by means of a frequency polygon or histogram, to see whether we are dealing with a bimodal, seriously skewed, or some other kind of distribution. And if the deviations from a normal distribution are excessive, we cannot legitimately make use of many of the more elaborate statistical tools discussed in subsequent chapters.

Exercises

DATA: *Aptitude Test Scores*

33 45 42 58 51 55 46 48 46 42 47 42 58 39 47 62 42 68 70 48
55 51 60 38 55 52 50 60 61 67 48 53 59 39 50 48 34 32 53 65
46 56 52 54 43 44 45 44 ($n = 48$)

1. Make a tally and frequency table from the above data, using as your lowest step 30–34.
2. Plot (on graph paper) a frequency polygon from this table.
3. Plot a histogram from this same table.
4. Is the distribution of scores approximately normal? Or is there a tendency toward bimodality, or toward skewness?
5–8. Repeat Exercises 1–4, using 30–32 as the bottom step and using both the simpler and the alternative designations for the steps.
9. Using the first 24 aptitude test scores above as your data and 30 as the lowest score which appears in your table, choose an appropriate step interval and make a tally and frequency table. Use the simpler designation for the steps.
10. Plot a histogram from the frequency table obtained in the preceding exercise.

11. Using the last 24 aptitude test scores above as your data, and the same steps used in Exercise 9, make a tally and frequency table.

12. Plot a frequency polygon from the frequency table obtained in the preceding exercise.

13. Plot a frequency polygon from the frequency table obtained in Exercise 9 and compare it with that obtained in Exercise 12. If any skewness is present, is it in the same direction? Does either frequency polygon show a tendency toward bimodality? Give a reason why you should not necessarily expect the two to be closely similar.

3
Measures of Central Tendency

We have just seen how the frequency table and the more vivid frequency polygon and histogram plotted from it give us a general idea of the way in which a given set of scores is distributed. The use of any one of these devices enables us to read some meaning into a disorganized jumble of scores (or other measures), such as those dealt with in the preceding chapter. Yet these devices, though essential for preliminary analysis, are relatively clumsy. We should like, among other things, some *single number* to represent the general level of performance (to continue our aptitude test illustration). There are three common devices for so expressing the center of concentration of scores in any set of data: the arithmetic mean, the crude mode, and the median. These *measures of central tendency*, which may be determined either for grouped data (as in the frequency table on

20   Measures of Central Tendency

page 10)or for ungrouped data (such as the unorganized set of scores from which the frequency table was derived), are discussed in the following sections.

1. The Mean

Though there are several means in common use, the arithmetic mean is the one most frequently employed in statistics. It is usually called simply *the mean*, and is nothing more than the familiar average. It is not necessary to explain how it is determined for *ungrouped data*. However, because of the important part it plays in later calculations, we should think of it in terms of a formula. If we let X stand for the individual scores, the most compact symbol for the mean of these scores is $\bar{X}$ (called "bar-X"). The formula for the mean is then

$$\bar{X} = \frac{\Sigma X}{n} \text{ (ungrouped data)} \tag{3.1}$$

where Σ (large sigma, the Greek symbol for "capital S") stands for "the sum of," and n (as above) for the total number of scores.

When we have a large number of scores to deal with, the calculation of the mean is a tedious job. However, when n is large we shall ordinarily group our data into a frequency table. Under these circumstances, with our *data grouped*, the mean may be calculated with much less effort and with but little loss of accuracy. The method is illustrated in Table 2. Here the scores in each step are represented by their respective midpoints (X).*

TABLE 2

Calculation of the Mean, the Crude Mode, the Median, and the Quartile Deviation from Grouped Data

(DATA FROM TABLE 1)

Step	Midpoint (X)	f	fX
65.5–68.5	67	1	67
62.5–65.5	64	2	128
59.5–62.5	61	4	244
56.5–59.5	58	4 $\atop{\displaystyle\frac{}{5}}$(31)	232
53.5–56.5	55		275

*The assumption that the midpoints are truly representative of all the scores in their respective steps does not introduce any important error when n is large and the distribution approximately normal.

TABLE 2 (continued)

Step	Midpoint (X)	f	fX
50.5–53.5	52	7 ⎫(19)	364
47.5–50.5	49	6 ⎭	294
44.5–47.5	46	5 ⎫	230
41.5–44.5	43	3 ⎬ (8)	129
38.5–41.5	40	2	80
35.5–38.5	37	1	37
32.5–35.5	34	2	68
		$n = 42$	$\Sigma fX = 2148$

A. *The Mean*

$$\bar{X} = \frac{\Sigma fX}{n} = \frac{2148}{42} = 51.1 \qquad (3.2)$$

B. *The Crude Mode*

The modal step is 50.5–53.5. The crude mode is the midpoint, 52.

C. *The Median (Mdn)*

$$\frac{n}{2} = \frac{42}{2} = 21 \text{ (``quota'')}^a$$

$$Mdn = 50.5 + (\tfrac{2}{6} \times 3) = 50.5 + 0.9 = 51.4$$
$$\uparrow$$
$$(step\ interval)$$

D. *The Quartile Deviation (Q)*

(1) *We first get* Q_1:

$$\frac{n}{4} = \frac{42}{4} = 10.5 \text{ (quota)}$$

$$Q_1 = 44.5 + \left(\frac{2.5}{5} \times 3\right) = 44.5 + 1.5 = 46.0$$
$$\uparrow$$
$$(step\ interval)$$

(2) *We next get* Q_3:

$$\frac{3n}{4} = \frac{3 \times 42}{4} = 31.5 \text{ (quota)}$$

$$Q_3 = 56.5 + \left(\frac{0.5}{4} \times 3\right) = 56.5 + 0.4 = 56.9$$
$$\uparrow$$
$$(step\ interval)$$

(3) *We then substitute in the formula:*

$$Q = \frac{Q_3 - Q_1}{2} = \frac{56.9 - 46.0}{2} = 5.5$$

---

[a]This is not a standard term, but it seems helpful.

The fX column is filled in with the products of f (the frequency) and X for each step. The sum of the fX column, ΣfX, is then substituted in the formula:

$$\bar{X} = \frac{\Sigma fX}{n} \text{ (grouped data)} \qquad (3.2)$$

The mean has several special virtues: it is usually the most reliable measure of central tendency (since it is based on the total number and the individual values of all the scores), and it is made use of in calculating fundamental measures of variability (see Chapter 4) and other important statistical measures. The weakness of the mean lies in the laboriousness of its calculation when n is large, and in the fact that it is less representative than the median when the array of scores contains a few cases which are markedly atypical (see section 4 below). In section 5 two *short methods* for calculating the mean are presented.

2. The Crude Mode

The crude mode,* for *ungrouped data*, is simply the most frequent score. Thus, for the scores 8 11 16 11 8 11 12, the mode is 11. This is the crudest of the measures of central tendency and is to be used for very rough estimates only. With *grouped data*, the midpoint of the modal step, or the step with the greatest frequency, is taken as the crude mode. In the case of the data in Table 2 the modal step is 50.5–53.5; the crude mode would therefore be 52. (In the case of a tie between two steps with the maximum frequency, a point halfway between the two midpoints would be chosen for the crude mode.) Determined from grouped data, the crude mode is still a rough measure. The ease with which it can be obtained is its principal recommendation.

3. The Median

The median (Mdn) is the midpoint of an array of scores; or, more elaborately, the point above which and below which 50%

---

*The crude mode is sometimes referred to as *the mode*. This name, however, should be reserved for the *mathematical mode*, which is more useful in theoretical work, far more difficult to calculate, and beyond the scope of this text.

of the scores fall. (The scores, of course, must be considered *in order of magnitude:* there would be no particular value in determining the midpoint in the purely geographical sense.) Its determination for both ungrouped and grouped data will be discussed.

A. *The determination of the median for ungrouped data* is illustrated for three typical cases below:

Case I: n Is Odd (9)

Scores (in order of magnitude): 6 9 10 11 *12* 13 15 17 20

Here, where n is odd, the median score is the middle score, 12. If we want the *median point*, we take the midpoint of this *median score*, which is 12.0, since the score covers the interval 11.5 to 12.5. This leaves 4.5 scores (50% of them) above the median and 4.5 scores below it.

Case II: n Is Even (10)

Scores (in order of magnitude): 3 6 9 10 11 ↓ 12 13 15 17 20
11.5

Here, where n is even, there are two middle scores, 11 and 12. The median is halfway between these, at 11.5, the point which is at the same time the upper limit of score 11 (which stands for the interval 10.5–11.5) and the lower limit of score 12 (11.5–12.5).

Case III: Several Scores Have the Same Value as the Mid-Score

Scores (in order of magnitude): 6 7 8 11 11 11 11 13 15

Here the middle score is 11. For most purposes it is sufficiently accurate simply to take this middle score as the *median score*, but, if we want an accurate determination of the *median point*, we may follow the procedure which is described below for use with grouped data, taking 1 as our step interval.

B. *The calculation of the median when the data have been grouped* into a frequency table is illustrated in Table 2. (The X and fX columns do not enter into this calculation.) It may at first seem confusing, but it is a strictly logical procedure and with a little practice becomes almost an automatic process. The definition, the point above which and below which 50% of the cases fall, suggests the first step: find 50% of the cases, or $n/2$. In this case $n/2 = 21$. Our problem now boils down to finding the point below which just 21 scores fall. (This will of necessity also be the point above which just 21 scores fall.) We count up from the bottom of the frequency column. Taking all the scores up to the upper end of step 47.5–50.5, or (what amounts to the same thing) to the lower end of step 50.5–53.5, we have 19. If we add all the cases in step 50.5–53.5, we reach 26 cases, which exceeds our "quota" of 21. The point below which just 21 cases fall must, therefore, lie somewhere in step 50.5–53.5. It will, then, equal 50.5 plus a certain correction. The amount to be added is determined by interpolation (as in using a table of logarithms). Having used 19 cases in reaching the point 50.5, we need 2 more to complete our quota. In other words, we need $2/7$ of the cases on step 50.5–53.5 to reach 21 cases. Since the step represents a score range of 3 (i.e., the step interval is 3), it is logical to take $2/7$ of 3 as our correction to add to 50.5. Hence, the median is $50.5 + (2/7 \times 3) = 50.5 + 0.9 = 51.4$. (Anyone with sufficient curiosity may check this value by working down from the top to the point *above* which just 21 cases fall. The result will be the same.)

Though it may be hard to believe at first, the median is usually easier to calculate than the mean, especially when n is large and when a frequency table has already been worked out, as is often the case. Another reason for using it at certain times is given in the section following.

4. The Median Is Sometimes More Representative than the Mean

By this time there has doubtless arisen in some minds the question: Why do we need this third measure of central tendency, the median, when we already have the mean for reliable work and the crude mode for rough estimates? Two illustrations (involving the same principle) will make clear the need for the median in certain cases. The first is a somewhat fanciful one chosen in the interest of arithmetical simplification.

Let us suppose that a certain boys' club has agreed to take part in an informal interclub track meet. In order that tentative club handicaps may be worked out, the boys are asked to send in the average of their times on the 100-yard dash. Here are the scores for nine of the boys: 11 12 12 13 *13* 13 14 14 15. The mean for these nine cases is 13. Now there are two other boys in the club who cannot be expected to take part in the meet, a crippled boy and an exceedingly fat lad who detests all forms of exercise. But they are both good sports and wish to comply with the request of the interclub officers. So the crippled boy hobbles through the 100 yards in 25 seconds, and the fat boy rolls through it in 30 seconds. If we include these two atypical cases with the other nine, the mean jumps to 15.6. This is an increase of 2.6 seconds and gives a totally false impression of the ability of the club as a whole. Each of the two extra scores has had an influence on the mean approximately twice as great as that of any other score. Now, if we use the median for all 11 cases, rather than the mean, there will be little or no distortion; for each of the atypical cases will have no more influence on the median than any of the other scores: each score, whatever its size, counts once and only once. The median for all 11 cases is 13.3, which is nearly the same as the mean of the nine cases, those that really count when it comes to running.

A more important situation which calls for the median in preference to the mean is one in which we want to know the *typical* annual income for a large group of people. For example, the mean income for adult New Yorkers is lifted considerably above the median by the influence of a relatively small number of fabulous Wall Street incomes. The median value, however, is far more *representative*. In general, the median should be used when a few atypical cases of very large or very small value will distort the mean. Of course, in cases of this sort there is no law against using both the mean and the median for a more complete picture of the group. (When the frequency distribution is perfectly symmetrical, as in the case of a normal probability curve, the mean, median, and crude mode *all have the same value*.)

It is possible, however, for an unscrupulous person to give a wholly false impression by the use of the mean alone, while speaking the literal truth. It is the occurrence of such distortions which has called forth from some deluded victim the following rancorous epigram: "There are lies, damned lies, and statistics." But this is unfair to statistics. The truth of the matter has been well put: "It is not that figures lie, but that liars figure."

5. Short Methods for Calculating the Mean by Coding

There are several methods of reducing the size of scores when calculating the mean. They commonly involve reduction either by subtraction or by division or by both. Such score reduction is called *coding*. Two coding methods are illustrated in this section.

A. *The Mean of Large Ungrouped Scores Coded by Subtraction*

If we have to deal with a set of ungrouped scores that are so large as to be unwieldy, we may save time in calculating their mean by the simple device of subtracting *a convenient constant* from each score. We then go through the usual calculations for the mean with the simplified scores and add the constant to the result. This procedure may be expressed by a formula:

$$\bar{X} = \frac{\Sigma X'}{n} + c \qquad [\text{or } \bar{X} = \bar{X}' + c] \qquad (3.3)$$

where $X' = X - c$; that is, where the coded score X' equals the original score X reduced by the constant c. (The derivation of this formula is given in section 6.) The symbol $\bar{X}'$ stands for the mean of the X' scores and is equal to $\Sigma X'/n$.

The application of this formula is illustrated with simplified data in Table 3. We see that the mean calculated by this new formula (*3.3*) is identical with the mean calculated in the usual manner by the old formula (*3.1*). Of course the constant in this table could have been chosen so as to eliminate *minus signs*, but this is not necessary. The formula works either way. The main consideration is to *choose the constant for greatest convenience,* so that the subtraction can be done *easily in the head* and so that the resulting X' values will be small in comparison to the original X values. If this is not so, $\bar{X}'$ will be no easier to calculate than $\bar{X}$ and the method has no virtue. It would be something like determining the average weight of all the mongrel dogs on the street by cutting a pound off the end of each of their tails, getting the average weight of what was left

TABLE 3

Calculation of the Mean of Large Ungrouped Scores Coded by Subtraction

(IQ DATA)

Original Score (X)	Constant Subtracted from Each Score (c)	Coded Score (X')
105		5
125		25
95		−5
115		15
110	100	10
135		35
90		−10
120		20
130		30
100		0
$\Sigma X = 1125$	$n = 10$	$\Sigma X' = 125$

$$X' = X - c$$

$$\bar{X} = \frac{\Sigma X'}{n} + c = \frac{125}{10} + 100 = 112.5 \qquad (3.3)$$

As a check with the usual formula (*3.1*):

$$\bar{X} = \frac{\Sigma X}{n} = \frac{1125}{10} = 112.5$$

of the dogs, and adding a pound to the result! This would make it appropriate to refer to the procedure as "the curtailed score method."

B. *The Mean of Grouped Scores Coded by Subtraction and Division*

A parallel method may be used with a much more marked saving of time when we have a large number of grouped scores to deal with. Here we use a method which in effect simplifies the scores, not only by subtraction, but by division as well. But this is all done painlessly as illustrated in Table 4. In this table the data columns (those for *Step, Midpoint*, and frequency, *f*) are all taken from Table 2. The only new columns are those for the coded score (X') and for fX'. And these can be filled in by inspection. We first place a 0 in the X' column *at any point*. If we pick one of the central steps, the arithmetic will be simpler. We then label the corresponding midpoint in the X column c. Next we fill in the rest of the X' column by simply counting up and down from the 0 reference point, using minus signs for the steps below the 0. The fX' column is filled in by multiplying the f and X' values together for each step. If we now get the algebraic sum of this fX' column, $\Sigma fX'$, and substitute this sum in the formula below, we have obtained the value of the mean with almost no arithmetical labor.

$$\bar{X} = i\left(\frac{\Sigma fX'}{n}\right) + c \qquad [\text{or } \bar{X} = i\bar{X}' + c] \qquad (3.4)$$

where c may be any step midpoint we choose

$i = $ the step interval

$X' = \dfrac{X - c}{i}$ (by implication*; it does not have to be calculated), and

$\bar{X}' = $ the mean of the coded X' scores; namely, $\dfrac{\Sigma fX'}{n}$

We note that the value of the mean obtained by this short method is just the same as that obtained in Table 2 with consid-

*See text below.

TABLE 4

*Calculation of the Mean of Grouped Scores
Coded by Subtraction and Division*

(APTITUDE SCORE DATA FROM TABLES 1 AND 2)

Step	Midpoint (X)	f	Coded Score (X')	fX'
66–68	67	1	6	6
63–65	64	2	5	10
60–62	61	4	4	16
57–59	58	4	3	12
54–56	55	5	2	10
51–53	52	7	1	7
48–50	$c^a = 49$	6	0	0 (61)
45–47	46	5	−1	−5
42–44	43	3	−2	−6
39–41	40	2	−3	−6
36–38	37	1	−4	−4
33–35	34	2	−5	−10
		n = 42		(−31)

Step interval:
$i = 3$

$\Sigma fX' = 61 - 31 = 30$

$$\bar{X} = i\frac{\Sigma fX'}{n} + c \quad \left[\text{where } X' = \frac{X - c}{i} \right] \quad (3.4)$$

$$= 3 \cdot \frac{30}{42} + 49 = 2.14 + 49 = 51.14$$

This is the same value of $\bar{X}$ as that found by the much slower method of formula *3.2* in Table 2.

---

[a] c may be *any step midpoint*, but if it is chosen from one of the central steps, the fX' values will be much smaller.

---

erably greater arithmetical effort by formula *3.2*. This is so because formula *3.4* is actually derived from formula *3.2*. The rather simple derivation is given in the next section. It is based on the relationship between X and X'. Though we were not aware, perhaps, that we were subtracting the constant c from anything, or that we were doing any dividing at all, the procedure used in effect reduced the large X scores to the small and more manageable X' scores by both of these operations. In other words, $X' = (X - c)/i$. When we put a 0 in the X' column opposite midpoint 49, we in effect subtracted 49 from this and all the other midpoints. And when we used a 1-point step interval in the X' column, we in effect divided the 3-point interval in the X column by 3. We then got the mean of these greatly simplified scores, $\bar{X}' = \Sigma f X'/n$, and finally reversed the process of coding by using the formula for $\bar{X}$. This is why, though the formula looks more complicated, the arithmetic is much easier when we use this method.

6. Derivation of the Two Formulas for Calculating the Mean by Coding; Three Rules of Summation*

The simple derivations of formulas *3.3* and *3.4* discussed in the preceding section are given here for three reasons: (1) they help to explain these useful coding formulas; (2) they serve as an introduction to some elementary summation procedures (involving Σ); and (3) they will make more meaningful some of the formulas in later chapters and their applications.

A. *Three Summation Rules*

I. *The summation of a constant for n terms* is equal to n times the constant. For example: $\Sigma c = nc$.

II. *The summation of the product of a constant and a variable* is equal to the product of the constant and the summation of the variable. For example: $\Sigma cx = c\Sigma x$.

III. *The summation of a sum (or difference) of two or more terms* is equal to the sum (or difference) of the summations of the separate terms. For example:

$$\Sigma(x + y - z) = \Sigma x + \Sigma y - \Sigma z$$

or $$\Sigma\left(ax + \frac{y}{b} - z^2\right) = a\Sigma x + \frac{1}{b}\Sigma y - \Sigma z^2$$

*This section will be relatively meaningless to those who have omitted the previous section; it will be relatively painless to those who have not.

B. *Derivation of the Formula for the Mean of Ungrouped Scores Coded by Subtraction*

To show that
$$\bar{X} = \frac{\Sigma X'}{n} + c \qquad (3.3)$$

By definition the coded score $X' = X - c$
which may be written $X = X' + c$
Substituting this value of X in the usual formula for the mean (*3.1*), we get

$$\bar{X} = \frac{\Sigma X}{n} = \frac{\Sigma(X' + c)}{n}$$

By Rule III: $\quad \bar{X} = \dfrac{\Sigma X' + \Sigma c}{n} = \dfrac{\Sigma X'}{n} + \dfrac{\Sigma c}{n}$

By Rule I: $\quad \Sigma c = nc$, which substituted in the equation above gives us

$$\bar{X} = \frac{\Sigma X'}{n} + \frac{nc}{n} = \frac{\Sigma X'}{n} + c \qquad \text{Q.E.D.}$$

This may also be written $\bar{X} = \bar{X}' + c$, where $\bar{X}'$ is the mean of the coded scores.

C. *Derivation of the Formula for the Mean of Grouped Scores Coded by Subtraction and Division:*

To show that $\quad \bar{X} = i\left(\dfrac{\Sigma fX'}{n}\right) + c \qquad (3.4a)$

By definition the coded score $X' = \dfrac{X - c}{i}$

Solving for X we get

$$X = iX' + c$$

Substituting this value of X in the usual formula for the mean of *grouped* scores (*3.2*), we get

$$\bar{X} = \frac{\Sigma fX}{n} = \frac{\Sigma f(iX' + c)}{n}$$

By Rule III: $\quad \bar{X} = \dfrac{\Sigma fiX' + \Sigma fc}{n}$

By Rule II: $$\bar{X} = \frac{i\Sigma fX'}{n} + \frac{c\Sigma f}{n}$$

And since $\Sigma f = n$: $\bar{X} = i\left(\dfrac{\Sigma fX'}{n}\right) + c$ \hfill Q.E.D.

This may also be written $\bar{X} = i\bar{X}' + c$ \hfill (3.4b)

since $\Sigma fX'/n$ equals the mean of the coded grouped scores ($\bar{X}'$).

Exercises

DATA FOR EXERCISES 1–3: Use the aptitude test scores on p. 17.

1. Find the mean: (a) for ungrouped data; (b) for data grouped as in Exercise 1 in Chapter 2.

2. Find the crude mode: (a) for ungrouped data; (b) for data grouped as in Exercise 1 in Chapter 2.

3. Find the median: (a) for ungrouped data; (b) for data grouped as in Exercise 1 in Chapter 2.

4–6. Find the mean, the crude mode, and the median using the data grouped as in Table 2, *but* with the two top and two bottom steps omitted. ($n = 36$)

7. Using the first 10 aptitude test scores on p. 17, ungrouped, as your data, find the crude mode and the median.

8. Using the data of Exercise 7 above, ungrouped, calculate $\bar{X}$ by the method of Table 3.

9. Using the grouped data of Table 4, find $\bar{X}$ for the scores coded by both subtraction and division, *using $c = 52$*.

10. Repeat Exercise 9, using $c = 34$. Compare the result with that of Exercise 9. Explain.

11. For the grouped data of Table 4, *but* with the three bottom steps omitted ($n = 37$), find $\bar{X}$ for the scores coded by both subtraction and division.

4
Measures of Variability

Now that we know how to determine the general nature of any given frequency distribution and to present it in graphical form, and how to select the most representative measure of central tendency and work out its numerical value, our kit of statistical tools may seem to be complete. However, it is not difficult to show that an analysis which goes no further is inadequate and possibly even misleading. For example, imagine yourself to be a member of a mixed class in educational psychology. At the first meeting an intelligence test is given. A week later the instructor injudiciously announces that the mean score for the men was 115, and for the women, 110. (This is *not* a typical result.) Forthwith, a cheer arises from the male contingent. On meeting a certain lady member of the group after class, you may be tempted to say: "Aha, I told you so! The next time we have an

Measures of Variability

argument, perhaps you will be good enough to listen to me—we men are more intelligent."

This kind of conclusion (though usually expressed with more reserve) is unfortunately very common when two groups are compared on the basis of the means alone. The instructor in this case neglected to mention important facts concerning the variability, or "spread," of the scores about the mean in each group. He gave no indication of the *individual differences* within the two groups. Though his statement about the means was accurate enough, he failed to state that the scores in the men's group ranged from 100 to 130, and that they ranged from 85 to 135 in the women's group. Figure 5 gives an idea of the relationship of the distributions. With this information available, it becomes clear that the most significant point in the comparison is *not* the slight difference between the means but the *overlapping*. The shaded area in the figure indicates the extent of the overlapping. There is more of similarity than of difference. Furthermore, in this particular case (which, again, is *not* typical for sex comparisons), there is greater variability in the women's group; so that a very considerable number of women did better than the average man, and some women did even better than the best man. The fact that the mean for the men ($\bar{X}_M$) is higher than the mean for the women ($\bar{X}_W$) does not make it at all improbable that any individual woman should have a higher score than any individual man. *Individual differences within the groups are more important than the differences between the means.*

The inadequacy of the mean (or other measure of central

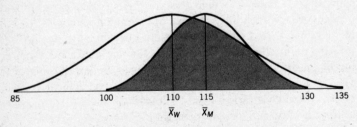

Figure 5. Overlapping distribution curves.

tendency) *alone* to characterize a group is clear. In addition some measure of variability is needed in order to bring out the extent of the individual differences within the group. Though the frequency polygon, or the histogram, gives us a rough idea of this, these graphs cannot be reproduced verbally. We need some single numerical value to express variability. Five such measures in common use will be discussed below.

1. The Range

We have already made use of the range in making our frequency table in Chapter 2. It is simply high score minus low score: Range = $H - L$. The range has about the same virtues and weaknesses as the crude mode. It is easy to get but none too valuable when you get it. Since it is based on two scores only (the top and the bottom), and since single scores are subject to large chance variations, a single atypical case at either end of the distribution would alter the value of the range considerably. We should like some more stable measure of variability which took into account more of the scores. In one way or another the remaining measures of variability do this.

2. The Mean Deviation (MD)

The mean deviation, MD, is a more reliable measure than the range. Every score in the distribution enters into its determination. It is usually defined as the mean of the deviations of the individual scores (or other measures) from the mean. It is determined for *ungrouped data* by the formula:

$$MD = \frac{\Sigma|x|}{n} \qquad (4.1)*$$

where x stands for the deviations of the individual scores from the mean, $(X - \bar{X})$, and $|x|$ stands for the *absolute* values of these deviations (that is, the value without regard to sign). The

*This formula may be presented in the equivalent form:
$MD = \Sigma|(X - \bar{X})|/n$; for by definition $x = (X - \bar{X})$.

other symbols are already familiar. The calculation of the mean deviation is illustrated with simplified ungrouped data in Table 5. The procedure is extremely simple. We first get the mean of the ungrouped scores (or other measures); in this case, 15. We next fill in the x column with deviations from the mean (from 15 here), and get the sum *ignoring the minus signs*. This sum divided by n gives us the value sought.

The full significance of MD as a measure of variability cannot be made clear without a little more background, which will be introduced in section 7; nevertheless the definition alone tells us something. Applying it to the illustration in Table 5, for example, we know that on the average the individual scores deviate from the mean score by two points only. We could now make a comparison between this group and some other with respect to uniformity of performance on the test in question.*
This would be a more meaningful comparison than one based on the score ranges alone.

3. The Standard Deviation (s) and the Variance (s^2)

Though the MD is a psychologically meaningful measure of deviation, its use of absolute values, which ignore minus signs, makes it a little disturbing to the mathematicians. This snubbing of the minus signs by the MD reduces its usefulness for algebraic manipulation. (If we retained the minus signs, however, Σx would become 0, which would be even more disturbing and would eliminate the usefulness of the MD altogether!) So two other more generally useful measures of variability are widely used, the standard deviation and the variance. These measures are closely related to each other, the standard deviation (s) being the square root of the variance (s^2); or, if you prefer, the variance is the square of the standard deviation. One or the other of these is used as the standard measure of variability in most statistical procedures. The calculation of these measures from simplified *ungrouped data* is illustrated in Table 5. The first two

---

*See Chapter 6 for a more accurate procedure when *different* tests are compared.

TABLE 5
*Calculation of the Mean Deviation (MD),
the Standard Deviation (s), and the Variance (s^2)
Illustrated for Ungrouped Data*

Scores (X)	Deviations from Mean $(x = X - \bar{X})$	x^2
19	4(.0)[a]	16(.00)[a]
16	1	1
15	0	0
11	−4	16
17	2	4
14	−1	1
18	3	9
13	−2	4
12	−3	9
15	0	0
$\Sigma X = 150$ $n = 10$	$\Sigma\lvert x \rvert = 20.0$ (Ignore signs; use absolute value.)	$\Sigma x^2 = 60.00$

$$\bar{X} = \frac{\Sigma X}{n} = \frac{150}{10} = 15.0^a \qquad (3.1)$$

$$MD = \frac{\Sigma\lvert x \rvert}{n} = \frac{20.0}{10} = 2.0 \qquad (4.1)$$

$$s^2 = \frac{\Sigma x^2}{n-1} = \frac{60.00}{9} = 6.67 \qquad (4.2)$$

$$s = \sqrt{\frac{\Sigma x^2}{n-1}} = \sqrt{\frac{60.00}{9}} = 2.6 \qquad (4.3)$$

[a] For the sake of clarity an example has been chosen in which the value of the mean is a whole number. Usually this will not be the case. Consequently, the values of x and x^2 will usually not come out as whole numbers either. It is seldom necessary, however, to express the values of $\bar{X}$ and x with an accuracy beyond the first decimal place. x^2 should be carried a bit further.

steps are the same as those required for the MD; namely, obtaining the mean of the scores and filling in the column of deviations from the mean, the x's. Next each x value is squared and the square is entered in the x^2 column. Finally, we get the sum of the x^2's (that is, Σx^2, often called "the sum of squares") and substitute in the formulas:

$$\text{Variance: } s^2 = \frac{\Sigma x^2}{n-1} \qquad (4.2)*$$

$$\text{Standard deviation: } s = \sqrt{\frac{\Sigma x^2}{n-1}} \qquad (4.3)*$$

We see from formula *4.2* that *the variance* is essentially the mean of the *squared deviations* from the mean (the x^2's), rather than the mean of the simple deviations from the mean (the x's), which is what the MD is. The squaring eliminates the minus signs with mathematical propriety, rather than by simply ignoring them. The variance, or its principal ingredient Σx^2, can, therefore, be employed in useful algebraic manipulations, as we shall see in sections 4 and 8 where other methods of calculation are presented. Furthermore, the variance is not encumbered by the radical sign, so it is easy to manipulate. But the variance is numerically much larger than the MD; so much of the time it is cut down to a more practical size by taking its square root, and the resulting standard deviation (s) is used instead. The significance of s and MD and the relationship between them are discussed further in section 7.

If anyone is bothered, as he should be, by the fact that in calculating the mean deviation we divide $|\Sigma x|$ by n, whereas in calculating the variance and the standard deviation we divide Σx^2 by $n - 1$, a little explanation is in order. The MD is primarily used in *descriptive statistics*, where the main interest is in describing the characteristics of a given set of scores, or other measures. The variance and the standard deviation, though they

---

*The symbol σ, sigma (the Greek equivalent of "small s"), is sometimes used in these formulas. The symbol s is used here to suggest that the formulas usually apply to *samples* of scores drawn by chance from a larger "population" of scores. We shall use σ elsewhere in connection with larger, or theoretically infinite, populations of scores.

are also used for descriptive purposes, are often used in the more sophisticated *inferential statistics*. In this latter case, we infer things, or make *estimates* about the characteristics of a larger collection, or *"population,"* of scores on the basis of a relatively small *sample* of scores taken *at random* (by chance) from the larger set. The *true* characteristics of the larger population (the mean, the variance, etc.) are called *parameters*. An *estimate* of a parameter based on one or more samples is called *a statistic*. Now, since different random samples from the same large population do not all show the same degree of variability, it is customary in calculating the variance and standard deviation to "correct for bias" by dividing Σx^2 by $n - 1$ instead of by n. The "bias" referred to is an *underestimate* of the population variability, since scores in a sample can hardly vary as much as scores in the parent population. Dividing by $n - 1$ gives a slightly larger estimate of the population variance and standard deviation than dividing by n, especially for small samples, which are more likely to give estimates that are too small.

4. Methods for Obtaining the Standard Deviation (s) and the Variance (s^2) by Calculating Machine and by Coding

In this section two important time-saving methods are presented for the calculation of the standard deviation and the variance. The first of these (in part A below) involves the formulas of section 3 converted into a form which is convenient for machine calculation, and which may in some cases save time without such help. The second method (B below) is a very helpful coding device for use without benefit of calculating machine. It is a simple extension of the method used for calculating the mean from coded scores which was given in Chapter 3, section 5. In order to avoid confusion from too many formulas, *the new methods will be presented in terms of formulas for "the sum of squares,"* Σx^2. *The variance* (s^2) *may then be determined by dividing by* $n - 1$, *and the standard deviation* (s) *will be the square root of the variance.*

A. (*Machine*) *Calculation of Variance* (s^2) *and Standard Deviation* (s) *from Original Scores Ungrouped*

In the preceding section we used a fundamental formula for variance based on *deviation scores* (x):

$$s^2 = \frac{\Sigma x^2}{n-1} \qquad (4.2)$$

where x represented the deviations of the original scores (X) from their mean ($\bar{X}$). The essential feature of this formula is Σx^2, "the sum of squares." As demonstrated in section 8 below, Σx^2 may be transformed algebraically into a form which makes use of the original scores (X) only; that is,

$$\Sigma x^2 = \Sigma X^2 - \frac{(\Sigma X)^2}{n} \qquad (4.4)$$

Using this formula, we may calculate the variance by simply dividing the result by $n - 1$ (and then get the standard deviation by taking the square root) without bothering with any deviation scores at all. The original score values, however, tend to be relatively large. Hence, a calculating machine is desirable, but not necessary. To illustrate the use of formula 4.4 let us apply it to the data of Table 5, page 37. This is done in Table 6. *Even without a machine*, but with the help of the Table of Squares at the end of this book, the calculation is very simple indeed. We observe that, though the new procedure requires no deviation scores, the results are the same as those obtained by the old formula (*4.2*), which did.

Formula 4.4 is useful for calculating both the mean and the variance in the same operation. A good *calculating machine* makes it possible to get the sums ΣX and ΣX^2 *simultaneously* for any number of scores. Getting the mean from ΣX and substituting Σx^2 in formula *4.2* to obtain s^2 is then "duck soup."

B. *Calculation of the Variance* (s^2) *and the Standard Deviation* (s) *from Coded Grouped Scores*

Since calculating machines are not only expensive but are a nuisance to operate on crowded buses and subways, it is helpful

Methods for Obtaining the Standard Deviation (s)

TABLE 6

Calculation of the Variance (s^2) and the Standard Deviation (s) from Original Scores Ungrouped
(DATA FROM TABLE 5)

Scores (X)	X^2
19	361
16	256
15	225
11	121
17	289
14	196
18	324
13	169
12	144
15	225
$\Sigma X = 150$	$\Sigma X^2 = 2310$
$n = 10$	

Sum of squares:
$$\Sigma x^2 = \Sigma X^2 - \frac{(\Sigma X)^2}{n} = 2310 - \frac{(150)^2}{10} \quad (4.4)$$
$$= 2310 - 2250 = 60$$

Variance:
$$s^2 = \frac{\Sigma x^2}{n-1} = \frac{60}{9} = 6.67^a \quad (4.2)$$

Standard deviation: $s = \sqrt{6.67} = 2.6^a$

[a] Values of s^2 and s are the same as those obtained by the method of Table 5.

to have available other quick methods of calculation. Fortunately, there is an answer to the itinerant statistician's prayer. As is shown in section 8 of this chapter, the formula for the sum of squares may be transformed in still another way. This is a tremendous time-saver for use with grouped data and without a calculating machine. It employs a coding device similar to the one used to get the mean in section 5 of Chapter 3. When our data are grouped in a frequency table the sum of squares is symbolized by $\Sigma f x^2$ (rather than by Σx^2). The formula for $\Sigma f x^2$ when the scores are also coded looks formidable, but it is not:

$$\Sigma f x^2 = i^2 \left[\Sigma f X'^2 - \frac{(\Sigma f X')^2}{n} \right] \qquad (4.5)$$

where f represents the frequencies, i is the step interval, and X' represents the deviation of the step midpoints from some chosen midpoint, *coded by division by* i. You will notice the similarity in the general form of this formula to formula 4.4. Except for the inclusion of the i^2 factor, the only difference is the substitution of fX' for X. This makes sense because here the scores are both coded and grouped. We have used the symbol X' for coded X before, and f is always involved when we work with grouped scores. The application of formula 4.5 is made clear in Table 7, which has the same form and same column headings as Table 4 used to obtain the mean from coded scores (p. 29). In both tables *any midpoint near the center* is chosen and a 0 is entered opposite it in the X' column. The X' and fX' columns are filled in by inspection, as indicated. The procedure illustrated in Table 7 merely requires the filling in of the additional fX'^2 column. This is most easily done by multiplying the entries in the X' column by the corresponding entries in the fX' column. We then enter the values of i, n, $\Sigma fX'$, and $\Sigma fX'^2$ in formula 4.5 and out comes the sum of squares, $\Sigma f x^2$. The variance (s^2) is then obtained by dividing by $n - 1$, as usual (formula 4.6). The standard deviation (s) is of course the square root of this.

Before we make use of the last column in Table 7 as a check on the accuracy of our calculations (see next section), we should notice some interesting features of the columns already used. Although X' represents the coded step midpoints, we do not need a column of midpoints in the table at all! And we don't really care what step midpoint is chosen for our 0 reference point! (It simplifies the arithmetic, however, to have it near the middle.) There is no constant (c) to add as there was in the formula for the mean based on coded scores (3.4). This is because variability, which our present formula deals with, is independent

TABLE 7

Calculation of the Variance (s^2) and the Standard Deviation (s) from Coded Grouped Data

Step	f	X'[a]	fX'	fX'^2	$[f(X' + 1)^2]$[b]
50–54	1	5	5	25	(36)
45–49	2	4	8	32	(50)
40–44	2	3	6	18	(32)
35–39	4	2	8	16	(36)
30–34	7	1	7 (34)	7	(28)
25–29	5	0	0	0	(5)
20–24	3	−1	−3	3	(0)
15–19	1	−2	−2	4	(1)
10–14	1	−3	−3(−8)	9	(4)
$i = 5$	Σ: 26 = n		26	114	[192][b]

$$\Sigma fx^2 = i^2 \left[\Sigma fX'^2 - \frac{(\Sigma fX')^2}{n} \right] \quad (4.5)$$

$$= 5^2 \left[114 - \frac{(26)^2}{26} \right] = 2200$$

Therefore

$$s^2 = \frac{\Sigma fx^2}{n-1} = \frac{2200}{25} = 88 \quad (4.6)$$

And

$$s = \sqrt{88} = 9.4$$

---

[a] The 0 in the X' column may be placed opposite *any* step, but preferably near the middle. We do not need the step midpoints (X) in this procedure at all, nor any subtraction constant c. This is because the variability is the same no matter where the mean is located. See derivation of the formula in section 8 for explanation.

[b] For use only in checking; see next page.

of the location of the mean. The coding here boils down to coding by division only. In the derivation of this formula in section 8 the disappearance of c is explained.

5. Accuracy of Calculations: Charlier's Check

The data in Table 7 for obtaining the variance from coded scores were somewhat simplified; so it was difficult to make a mistake in the calculations. For larger values of n mistakes are possible. There is, however, an ingenious but simple device known as *Charlier's check* which tells us at once if a mistake has been made in filling in the columns involving f and X' or in getting the sums Σf, $\Sigma fX'$, and $\Sigma fX'^2$. The check is based on the fact that:

$$(X' + 1)^2 = X'^2 + 2X' + 1$$

and, therefore

$$\Sigma f(X' + 1)^2 = \Sigma fX'^2 + 2\Sigma fX' + \Sigma f \quad \text{(Charlier's check)}$$

where $\Sigma f = n$. It will be seen at once that the sums on the right side of this equation are the three sums which enter into our Σfx^2 formula (4.5). All we need, in order to check their accuracy in one fell swoop, is the sum on the left hand side, $\Sigma f(X' + 1)^2$. It is usually possible to write in a $f(X' + 1)^2$ column for such a table as Table 7 *by inspection* and then get the sum with very little effort. For example, the entry in the first row in the supplementary column was simply: $f(X' + 1)^2 = 1 \times (5 + 1)^2 = 36$. The other values in this column are almost as easy to work out in the head. Then entering the four sums in Charlier's check, we get

$$192 = 114 + 2 \times 26 + 26 = 192$$

Now we can relax!

6. The Quartile Deviation (Q)

The quartile deviation (Q) is commonly used in conjunction with the median. It may be defined as the average amount by which the quartile points, Q_1 and Q_3, deviate from the median. Q_1 and Q_3 are, respectively, the 25th and 75th percentile points; that is, Q_1 is the point below which just 25% of the scores (or other measures) fall, and Q_3 is the point below which just 75% of the measures fall. These points, together with the median (the 50th percentile point), divide the total distribution of measures into four equal parts, or quartiles; hence, the name "quartile points" for Q_1 and Q_3. The quartile deviation is sometimes known by the rather ponderous title "semi-interquartile range." Though such top-heavy labels usually contribute more to confusion than to clarity, in this case the name helps us to remember the formula:

$$Q = \frac{Q_3 - Q_1}{2} \qquad (4.7)$$

for $Q_3 - Q_1$ is obviously the interquartile range, and the 2 obviously makes it "semi." (This also gives us the mean of the deviations of the quartile points from the median.)

The application of this formula, illustrated in Table 2 (p. 21), involves the same general procedure as that required in the determination of the median. In getting Q_1, the first step, we work out our quota: $n/4 = 10.5$. We wish to find the point below which just 25%, or 10.5, of the measures fall. Counting up from the bottom, we find that the first four steps contribute 8 scores. If we add on all 5 cases in the next step (44.5–47.5), we have exceeded our quota. Hence, Q_1 must lie somewhere in the step 44.5–47.5. It must therefore equal 44.5 plus a certain correction. The correction is determined by interpolation, as in finding the median. This is fully illustrated in the table. By a similar procedure we get Q_3, using this time $3n/4$ as our quota. The final step is substitution in the formula.

The significance of Q is rather easy to see from the method of its determination, or from the definition. Assuming that a given distribution, for which we have worked out the values of Mdn and Q, is approximately normal, and therefore symmetrical, we are able to state the limits within which the middle 50% of the cases actually fall: we simply subtract Q from Mdn to find the lower limit, and add Q to Mdn to find the upper limit.* The knowledge of these limits gives us a concrete idea of just how concentrated about the median, or how dispersed about it, the scores are. Furthermore, we can now interpret our results

*If the distribution is skewed, the limits $Mdn \pm Q$ will also contain 50% of the cases, but they will not be located in the center of the baseline.

in terms of *probability*. We can say that the chances are 50 in 100 that any score taken at random from the group will fall within the limits $Mdn - Q$ and $Mdn + Q$.

The quartile deviation may be derived from ungrouped data by the same formula, but it will be found more useful with grouped data. Though in general somewhat less reliable than MD and s, it is the appropriate measure of variability to use with the median.

7. The Interrelation of Q, MD, and s

These three measures of variability all do the same general job. Each enables us to state the limits within which a certain percentage of the cases in the central region of a given distribution actually fall; thus each gives a measure of the spread of the cases about the central point. If derived from a normal frequency distribution, these measures have a fixed relation to each other. This is suggested in Figure 4 (p. 13). For such a normal probability curve, the mean and the median are exactly at the middle of the distribution. Q, when laid off above and below this middle point, sets the limits for the middle 50% of the cases. MD is a somewhat larger value than Q and, when laid off above and below the mean, sets the limits for approximately the middle 57% of the distribution. The largest of these measures is s, which, when used in a similar fashion, indicates the limits for the middle 68% of the cases, approximately.

There is no obvious explanation of these last two percentages. They are simply based on the characteristics of the normal probability curve. But this need not concern us further here. The interpretation is what we care about. *In terms of probability*, we can say that the chances are about 57 in 100 that an individual selected at random from the group which produced the scores obtained a score (X) between the limits $\bar{X} - 1MD$ and $\bar{X} + 1MD$. Similarly, the chances are about 68 in 100 that the score of an individual selected at random from the group will fall between the limits $\bar{X} - 1s$ and $\bar{X} + 1s$.

It is also interesting to observe, though this cannot be demonstrated here, that for a normal distribution *approximately*

95% of the cases fall between the limits $\bar{X} - 2s$ and $\bar{X} + 2s$, and *approximately* 100% of the cases fall between the limits $\bar{X} - 3s$ and $\bar{X} + 3s$. (See Figure 4, p. 13, for an illustration of this.) These facts may also be expressed in terms of probability: the chances are about 95 in 100 that a score taken at random will fall between the first set of limits, and the chances are nearly 100 in 100 that a random score will fall between the second set of limits. (See Chapter 7 for an explanation of this.)

8. Derivation of Special Formulas for the Sum of Squares Required for the Variance and Standard Deviation*

A. *Derivation of Σx^2 Formula used in (Machine) Calculation of Variance (s^2) and Standard Deviation (s) from Original Scores Ungrouped*

To show that

$$\Sigma x^2 = \Sigma X^2 - \frac{(\Sigma X)^2}{n} \qquad (4.4)$$

By definition

$$x = (X - \bar{X})$$

Therefore

$$\Sigma x^2 = \Sigma (X - \bar{X})^2 = \Sigma (X^2 - 2X\bar{X} + \bar{X}^2)$$

By summation rule III (p. 30) this becomes

$$\Sigma x^2 = \Sigma X^2 - \Sigma 2X\bar{X} + \Sigma \bar{X}^2$$

By summation rules I and II (p. 30) this then becomes

$$\Sigma x^2 = \Sigma X^2 - 2\bar{X}\Sigma X + n\bar{X}^2$$

Then, since $\bar{X} = \dfrac{\Sigma X}{n}$, we get by substitution

$$\Sigma x^2 = \Sigma X^2 - \frac{2(\Sigma X)^2}{n} + \frac{(\Sigma X)^2}{n}$$

*If section 4 has been omitted, it is a pity; this section will be superfluous. These derivations, though not essential to the use of the formulas, are not difficult. They may illuminate or even entertain.

Therefore

$$\Sigma x^2 = \Sigma X^2 - \frac{(\Sigma X)^2}{n} \qquad \text{Q.E.D.}$$

B. *Derivation of Σfx^2 Formula for Obtaining Variance (s^2) and Standard Deviation (s) from Coded Grouped Scores*

To show that the sum of squares for scores both grouped and coded is

$$\Sigma fx^2 = i^2\left[\Sigma fX'^2 - \frac{(\Sigma fX')^2}{n}\right] \qquad (4.5)$$

where i is the step interval and X' represents the *coded* values of the step midpoints expressed as deviations from some chosen midpoint (c) corresponding to $X' = 0$. In other words, $X' = (X - c)/i$. This is the formula for X' we used when we were calculating the mean from grouped scores coded by subtraction and division (p. 28). From this we got

$$X = iX' + c \qquad (1)$$

The corresponding formula for the mean was

$$\bar{X} = i\bar{X}' + c \qquad (2)$$

Subtracting (*2*) from (*1*), we get an expression for x in terms of the *coded* scores:

$$x = (X - \bar{X}) = (iX' + c) - (i\bar{X}' + c)$$

which reduces to

$$x = i(X' - \bar{X}') \qquad (3)$$

(In passing we notice that the constant c has disappeared from the expression for x. This is because x is a measure of *deviation from the mean*. It is completely independent of the location of the mean. For the same reason c does not appear in our final expression for Σfx^2.)

Now substituting the value of x from equation (*3*) in Σfx^2 and applying our rules of summation, we get

$$\Sigma fx^2 = i^2 \Sigma f(X' - \bar{X}')^2 = i^2 \Sigma f(X'^2 - 2X'\bar{X}' + \bar{X}'^2)$$
$$= i^2(\Sigma fX'^2 - 2\bar{X}'\Sigma fX' + \bar{X}'^2 \Sigma f)$$

Substituting the identities $\Sigma f = n$ and $\bar{X}' = \dfrac{\Sigma fX'}{n}$, we arrive at our intended destination

$$\Sigma fx^2 = i^2 \left[\Sigma fX'^2 - \frac{(\Sigma fX')^2}{n} \right]$$ Q.E.D.

Exercises

DATA:

Set A (Ungrouped)

72 51 46 58 48 47 58 47
70 51 38 52 60 67 53 39
($n = 16$)

Set B (Grouped)

Step	f
70–74	1
65–69	3
60–64	5
55–59	9
50–54	10
45–49	8
40–44	5
35–39	3
30–34	2

$n = 46$

1. Find the range for Set A.

2. Find MD for Set A.

3. Find Q for Set B.

4. Find the variance and the standard deviation for Set A using the formula for deviation scores.

5. Find the variance and the standard deviation for Set A using the formula for original scores ungrouped.

6. Find the variance and the standard deviation for Set B using the formula for coded grouped data.

7. Run a Charlier check on the calculations in Exercise 6.

8–11. Repeat Exercises 1, 2, 4, and 5 using the first 10 scores of Set A only.

12–14. Repeat Exercises 3, 6, and 7 using Set B, *but omitting* the top step and the two bottom steps. ($n = 40$)

15. *Review of Symbols Used in Chapters 3 and 4.* Match the symbol with the name or concept, putting the letter which is beside the symbol in the ( ) beside the proper name or concept. (See next page.)

Measures of Variability

Symbol		Name or Concept
a. MD	h. s^2	( ) variance ( ) mean
b. s	i. c	( ) deviation score ( ) standard deviation
c. $\bar{X}$ or $\bar{X}'$	j. Σx^2	( ) number in sample ( ) sum of squares
d. X'	k. f	( ) subtraction constant ( ) the sum of
e. x	l. Σ	( ) mean deviation ( ) original score, or step midpoint
f. X	m. n	( ) coded score ( ) frequency
g. i		( ) step interval

16. *Review of Formulas Introduced in Chapters 3 and 4.* Match the formula with the name or concept, putting the letter which is beside the formula in the ( ) beside the proper name or concept.

Formula   *Name or Concept*

a. $\dfrac{\Sigma|x|}{n}$ ( ) mean, grouped scores coded by subtraction and division

b. $\sqrt{\dfrac{\Sigma x^2}{n-1}}$ ( ) variance

c. $i\left(\dfrac{\Sigma fX'}{n}\right) + c$ ( ) mean, ungrouped scores coded by subtraction only

d. $\dfrac{\Sigma x^2}{n-1}$ ( ) sum of squares, scores grouped and coded by division

e. $\dfrac{\Sigma X'}{n} + c$ ( ) mean, scores grouped but not coded

f. $\dfrac{Q_3 - Q_1}{2}$ ( ) standard deviation

g. $i\bar{X}' + c$ ( ) mean, ungrouped original scores

h. $i^2\left[\Sigma fX'^2 - \dfrac{(\Sigma fX')^2}{n}\right]$ ( ) sum of squares, ungrouped original scores

i. $\dfrac{\Sigma fX}{n}$ ( ) quartile deviation

j. $\Sigma X^2 - \dfrac{(\Sigma X)^2}{n}$ ( ) mean deviation

k. $\dfrac{\Sigma X}{n}$

Note: In one instance the same name or concept applies to two formulas.

5
The Use of Norms and Grading "On the Curve"

1. The Significance of Scores Is Relative

Test scores have no *absolute* significance. It is obvious, for example, that a score of 50 on a test of only 50 items may have different implications from a score of 50 on a test of 100 items. And even if we express a subject's score in percentage terms, we may have said very little about his ability; for if, on a given test, all the subjects get 100% of the items correct, there is no distinction in a perfect score. On the other hand, on a more difficult test, a score of only 70% might indicate unusual ability. A score takes on significance only by comparison with other scores. Its significance is *relative*, not absolute.

These considerations apply as well to measures other than test scores. For example, a man is not considered tall simply because he is six feet four. (He might be a runt among Martians.)

He is considered tall because the majority of people are shorter.

Now let us consider some of the common methods of indicating relative standing in an array of test scores or other measures.

2. Quartile, Decile, and Percentile Norms

The significance of any test score is determined by comparing it with some *norm*, or standard (from the Latin *norma*, rule). The measures of central tendency discussed in Chapter 3 are in themselves rough norms; they give us the typical result (if derived from a large and unselected sample). We know something about an individual's performance if we merely know whether his score is above, at, or below the mean, or the median. We get a less rough idea of a subject's standing if we divide the distribution of scores into *quartiles* (or quarters), by means of the median and the quartile points, Q_1 and Q_3 (see Table 2). We can then place him in the upper quartile, lower quartile, and so on. If we wish to locate a score (or other measure) more precisely, we can break up the frequency distribution into *deciles* (10ths), or even into *percentiles** (100ths).

The first decile (lowest 10th) is bounded at its upper end by the 10th percentile point (P_{10}), the point below which just 10% of the scores fall. The second decile lies between the 10th and 20th percentile points (P_{10} and P_{20}), and so on. These percentile points, and also the ones which break up the distribution into smaller subdivisions (e.g., the 17th, 79th, and so on), are calculated in the same manner that Q_1 and Q_3 (i.e., P_{25} and P_{75}) were calculated. For example, if we wish to calculate P_{17}, say, we first determine our quota, $17n/100$. Then we count up from the lower end of the frequency column to the step in which our quota lies and, by interpolation, get the precise location within the step (see Table 2). Of course it would be silly to get percentile norms when n is much less than 100, or even decile norms when n is small. The choice of appropriate norms will depend in part on the size of the group.

*The term *centile* is preferred by some as a substitute for *percentile*.

3. Grading "On the Curve"

Another extension of the techniques discussed in earlier chapters, one with which the destiny of the student is closely linked, is grading "on the curve." This is a method for converting numerical scores on examinations (usually objective examinations) into letter grades. It makes use of the principle of relative standing discussed in section 1. A "C" (or other mediocre grade) is assigned to scores near the center of the distribution, a "B" to those some distance above, and an "A" to those very far above. The scores below average are treated in similar fashion. The determination of the *boundary scores* for five grade groups, A to E, is illustrated in Figure 6.

This procedure makes use of the median and Q (here assumed to be 50.5 and 6, respectively), though the mean and s may be employed if greater accuracy is desired. Since in most practical problems nearly all the scores fall within the limits $Mdn - 3.75Q$ and $Mdn + 3.75Q$, the assumption is made that the entire base line of the distribution curve is approximately $7.5Q$ long.* Now if we wish to divide this into five equal parts for our five grade groups, we simply divide $7.5Q$ by 5 (which

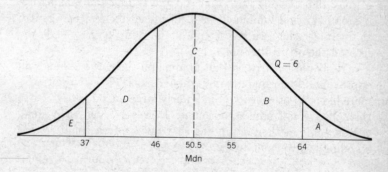

Figure 6. Grading "on the curve."

*For a normal distribution curve, 98.8% of the cases lie between the limits $\bar{X} - 2.5s$ and $\bar{X} + 2.5s$ (see Chapter 7), which are equivalent to the limits $Mdn - 3.75Q$ and $Mdn + 3.75Q$.

gives us $1.5Q$) to get the length of the "grade base line"—in this case, $1.5Q = 1.5 \times 6 = 9.0$. The C group is symmetrically placed with respect to the median. Its limits are therefore determined by laying off half of this value below and above the median: $50.5 - 4.5$ and $50.5 + 4.5$; i.e., 46.0 to 55.0. Adding 9.0 to 55.0, we get the upper limit of the B group, 64.0. This is also the lower limit of the A group. Similarly we get the boundary scores for the grades below C.* (Scores which actually fall on a boundary between two grade groups must be arbitrarily assigned to one or the other of the adjoining groups.) A similar procedure would be followed if we wished to divide the range of scores into the grade groups A to F, but the total base line would be divided into six rather than five groups; thus, the grade base line would be $7.5Q \div 6 = 1.25Q$, rather than $1.5Q$. And in this case, the median would be the boundary line between the C and D groups.

Lest the student be concerned with the possibility of the instructor's becoming mechanical in his methods and abdicating his customary power of arbitrarily assigning grades, it should be pointed out that, at the discretion of the instructor, the entire scale may be moved up or down. Thus, if he be in a genial mood, he may slide it down, crowding out the E area and making the A area large and luminous. Or, if his mood be less than genial, as is usually the case at the end of the term, he may slide the scale in the other direction.

It should be stressed that grading on the curve assumes an approximately normal distribution of scores. This assumption is more likely to prove correct in a fairly large introductory course than in a small advanced course. However, even in a small course, some method of grading that takes into account *relative* standing is always in order.

---

*When working with the mean and the standard deviation, the entire base line may be taken as $5s$ in most practical problems. The grade base line for five grade groups then becomes $1s$. The rest of the procedure is the same. These procedures, when applied to a normal distribution of scores, give approximately 7% A's, 24% B's, 38% C's, 24% D's, and 7% E's.

Exercises

DATA: *Scores on a history examination*

Step	f
70–74	8
65–69	12
60–64	20
55–59	34
50–54	46
45–49	34
40–44	22
35–39	14
30–34	10
	$n = 200$

1. Find the 1st, 5th, and 9th decile points (i.e., P_{10}, P_{50}, and P_{90}).
2. Find P_{37}, P_{52}, and P_{99}.
3. Find the boundary values (to the nearest whole score) for 5 grade groups, A to E.
4. Repeat Exercise 3, using 6 grade groups of equal score range, A to F.
5. Using the same steps as in the table above and a frequency for each step which is just one half of the value listed, find P_{15}, P_{25}, and P_{75}.
6. For the data of Exercise 5, find the score range between P_{10} and P_{90}. How does this compare with the corresponding value for the original frequency table? Explain.
7. Repeat Exercise 3 with the data of Exercise 5.
8. Repeat Exercise 3 using a frequency table which is the same as the original except that the f column is inverted.
9. Repeat Exercise 4 using the data of Exercise 8.
10. Find P_{20}, P_{80}, and P_{65} for the data of Exercise 8.

6
Standard Scores for Comparing and Combining Test Results*

In both psychology and education we are often faced with the problem of comparing an individual's scores on two or more tests, or with the related problem of combining such scores into a single measure. These problems cannot always be handled by simple, homemade techniques, but require the application and extension of some of the statistical tools developed in earlier chapters. For example, how would you compare John Jones's

---

*Throughout this chapter and part of the next the symbol σ is used in place of s to represent *the standard deviation*. This is partly to make this symbol more familiar, since it is widely used in much of the literature. And it is used partly because much of this chapter and the next deals with theoretical normal distributions, where the σ notation is almost always used. (See note on p. 38, where s is first introduced as a symbol for the standard deviation of *a sample* of scores from a larger "population" of scores.)

height in inches with his weight in pounds? Is John taller than he is heavy? Put in this way, the question is meaningless. It takes on meaning, however, if we introduce height and weight norms for boys of John's age. It is a meaningful question to ask how much taller or shorter John is than the mean height for his age. The corresponding question with respect to his weight is also meaningful. Suppose the answers are "10 inches taller" and "10 pounds heavier." Are we now justified in saying that John is as much taller than boys of his age as he is heavier? No, because inches and pounds are not directly comparable. We cannot assume that a positive deviation of 10 inches in the distribution curve for height corresponds to a positive deviation of 10 pounds in the weight curve. On the contrary, 10 units might represent nearly all of the range above the mean in one scale and only a small fraction of this range in the other. (And this objection would also hold even if our units were nominally the same in two different distributions, both inches, say, if the variabilities were not equal.) But, if we could somehow express John's height and weight in terms of their *relative* positions in the height and weight curves, we would be able to make a meaningful comparison. If we could indicate by single numbers of some sort just how much his scores differed from their respective means, not in terms of inches or pounds, but in terms of characteristics of the distribution curves themselves, we could then compare them, or, if we wished, combine them into a rough index of physical development. We can do this with the aid of standard scores or the other devices discussed below.

1. Standard Scores (z Scores)

Standard scores travel under many aliases: sigma scores, z scores, etc.; but they have no reason to be ashamed of themselves. On the contrary, they possess several virtues. In particular, they meet the needs discussed above: they may serve as the *common units* into which other scores may be converted, even though the original units be as different as cabbages and kings. Furthermore, a standard score is of such a nature that it expresses a subject's *position in a given distribution both with respect to the mean and with respect to the variability.* It is thus relatively free from ambiguity.

Any score may be converted into a standard score, provided the mean and σ for the distribution of which it is a part are known. A commonly used type of standard score is the z score. This is simply a score whose deviation from the mean is expressed in σ units. Reference to Figure 7 will make this clear. Here the original scores may be converted into z scores by inspection, because the numerical factors have been simplified, but the following formula is usually more convenient:

$$z = \frac{X - \bar{X}}{\sigma}, \text{ or simply } z = \frac{x}{\sigma}$$

where z stands for the standard score, X for the original or "raw" score, $\bar{X}$ for the original mean, x for the difference $X - \bar{X}$ (i.e., the deviation of the score from the mean), and σ for the original standard deviation. The formula is given in the $X - \bar{X}$ form, as well as in the simple x form, in order to eliminate any doubt about the proper algebraic sign of the answer.

To illustrate the use of this formula, let us assume that our friend John gets a score of 40 in a geography test which has a mean of 64 and a σ of 15. His z score will then be:

$$z = \frac{40 - 64}{15} = \frac{-24}{15} = -1.6$$

Let us further assume that his z scores for two other subjects are calculated: for history, $+1.6$; and for grammar, $+0.8$. We

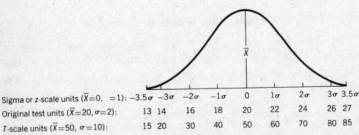

Figure 7. Test scores expressed in two different standard score equivalents: z scores and T scores.

could then make a direct comparison of the various pairs of scores. We could say, for example, that he is as much below his class in geography as he is above it in history; or that he is twice as far above the mean in history as he is in grammar.*
And, if we wish to combine his three scores, giving equal weight (or emphasis) to each, we can do this by simply averaging them, taking due account of the algebraic signs, thus: mean z score $= (-1.6 + 1.6 + 0.8) \div 3 = 0.27$. Or, if for any reason we wish to give special weight to some of the tests, we can multiply the z scores by the desired constants before combining them. It should be pointed out that these procedures for comparing or combining scores are not valid unless the scores come from the same group of subjects.

(Another important application of sigma scores appears in Chapter 7.)

2. T Scores

Standard scores expressed in z scale units are sometimes a little awkward to handle in practice, because of the common occurrence of negative signs and decimal values of sigma. As can be inferred from an inspection of Figure 7, a z scale is a normal distribution with $\bar{z} = 0$ and $\sigma = 1$. The principles which underlie its construction and use are in no way altered if the negative signs are eliminated by simply adding some constant. Nor is any principle violated if, in order to eliminate decimal values of the units, we enlarge the scale. Several such modifications of the z scale have been suggested. One in common use is the T scale introduced by McCall. The T scale is essentially a distribution with $\bar{T} = 50$ and $\sigma = 10$. Thus T scores are really standard scores derived from z scores by multiplying z by 10 and adding 50. This being the case, the formula used for calculating z scores can be readily modified into a formula for T scores:

$$T = 50 + 10z = 50 + \frac{10}{\sigma}(X - \bar{X})$$

*Unfortunately being "twice as far above the mean" cannot be simply interpreted as indicating success in work just twice as difficult. The interpretation of sigma scores is a more involved matter which should take into account the probability concepts dealt with in the next chapter.

where T stands for the T score, X for the original or "raw" score, $\bar{X}$ for the original mean, σ for the original standard deviation. If a single T score is to be calculated, this formula is more convenient; but if several scores are to be converted, the task may be simplified by an equivalent formula

$$T = K + RX$$

where R is the ratio $10/\sigma$ and K is the constant $50 - \bar{X}R$. The procedure is illustrated in Table 8.

TABLE 8

Converting Test Scores into Equivalent T Scores

Example A: A Single T Score

(When only one score is to be converted, the formula used below is the more convenient.)

Given: $\bar{X} = 40$, $\sigma = 5$, and $X = 55$

Then: $T = 50 + \dfrac{10}{\sigma}(X - \bar{X}) = 50 + \dfrac{10}{5}(55 - 40) = 80$

Example B: Several T Scores

(When several scores are to be converted from the same distribution, the alternative formula used below is the more convenient.)

Given: $\bar{X} = 60$, $\sigma = 3$, $X_1 = 66$, and $X_2 = 51$

Then: $R = \dfrac{10}{\sigma} = \dfrac{10}{3} = 3.33$

And: $K = 50 - \bar{X}R = 50 - 60 \times 3.33 = -150$

Hence: $T_1 = K + RX_1 = -150 + 3.33 \times 66 = 70$

Similarly: $T_2 = K + RX_2 = -150 + 3.33 \times 51 = 20$

NOTE: Scores on this scale are usually expressed without decimals; i.e., they are given correct to the nearest whole number.

The procedure just described is adequate for most purposes, but if we were doing an elaborate job of standardizing some psychological or educational test for publication, we should probably make use of *"normalized z scores."* "Normalizing"

z scores is a refined form of torture whereby an obtained distribution of scores, which may be somewhat skewed, is forced into a normal mold. The procedure is comparable to that used by the ancient robber Procrustes, who cut or stretched his hapless victims, however tall or short, to fit a bed of fixed length. This is beyond the scope and kindly intent of this short text.

3. Percentile Rank and Order of Merit Equivalents

We have just dealt with two refined methods of comparing and combining scores from different tests. Two more methods may be mentioned for use when less accuracy is required.

The first of these makes use of *percentile ranks* for the score equivalents. Unless tables for converting the scores from the various tests into percentile ranks are available, the method has neither great reliability nor convenience to recommend it. But if the scores from the different tests can be readily expressed in these terms, meaningful comparisons may be made directly. However, percentile scores cannot with complete propriety be combined by averaging; for averaging assumes scale units of uniform length, and this is not the case for the percentile scale.* The error introduced by averaging percentiles is considerably less if they lie between P_{25} and P_{75}, and is in other cases not so great as to rule out the method when moderate accuracy is all that is called for.

The second of these methods is based on *order of merit rankings*. It is the simplest but the crudest method of comparing or combining scores from different tests and should be used only when the number of subjects who have taken the tests is small. This method is illustrated for two sets of test scores in Table 9.

Once the scores have been ranked in order of merit, as in the 4th and 5th columns of this table, direct and meaningful comparisons may be made. For example, Subject 1 is 8th in a group

---

*This may be seen by reference to Figure 4, p. 13: the distance from the median to $1Q$ (i.e., from P_{50} to P_{75}) is less than one third the distance from $1Q$ to the upper limit of the distribution curve (i.e., from P_{75} to P_{100}), and not just equal to it as it would be if the percentile units were of uniform length.

TABLE 9

Combining Test Scores by Averaging Order of Merit Rankings

Subject's Number	Score in Test A	Score in Test B	Order of Merit Ranking[a]		Order of Merit of Combined Tests	
			Test A	Test B	Average Ranking	Reranked Average
1	10	23	8	7	7.5	8
2	18	27	1	2.5[a]	1.75	1
3	16	27	3	2.5[a]	2.75	2
4	13	32	6	1	3.5	3
5	12	25	7	5	6	7
6	17	18	2	8	5	5
7	15	26	4	4	4	4
8	14	24	5	6	5.5	6

[a]This method of ranking scores is explained in Chapter 13, section 1.

of 8 subjects on Test A and 7th on Test B; Subject 2 is 1st on Test A and tied with Subject 3 for 2nd and 3rd place on Test B. The last column gives the order of merit rankings of the eight subjects for the combined tests. The obvious defects of this method are that it takes no account of irregular gaps between successive scores and no account of the variability and form of the distributions. And it should be pointed out again, as it was in section 1, that none of these procedures for comparing or combining scores is valid unless the scores come from the same group of subjects.

Exercises

DATA: *Mean, σ and Scores of Three Students on Three Achievement Tests*

Arithmetic: $\bar{X} = 62$, $\sigma = 5$, $X_1 = 67$, $X_2 = 50$, $X_3 = 49$
History: $\bar{X} = 76$, $\sigma = 7.5$, $X_1 = 82$, $X_2 = 57$, $X_3 = 93$
Reading: $\bar{X} = 59$, $\sigma = 6$, $X_1 = 72$, $X_2 = 54$, $X_3 = 48$

1. What is the mean z score for Student 1 on all three tests?

2. What is the mean T score for Student 2 on all three tests?

3. By means of T scores, show whether X_3 in arithmetic is a better score than X_3 in reading.

4. By means of T scores, find which student has the best average performance in all three tests.

5. Convert the scores on Test A in Table 9 into T scores, using $\bar{X} = 14$ and $\sigma = 2$.

6. Convert the scores on Test B in Table 9 into T scores, using $\bar{X} = 25$ and $\sigma = 3$.

7. Combine the two sets of T scores obtained in Exercises 5 and 6 and arrange them in order of magnitude. Is the order the same as that obtained by the method of Table 9? Discuss.

7
Probabilities Determined from a Normal Distribution, the z Distribution; Confidence Limits for the Mean

We have been rather casually introduced to normal distribution curves in earlier chapters. There is a whole family of curves that satisfy the complex mathematical equation for a normal curve. We were usually referring to a particular form of normal distribution, based on standard or z scores, called the *z distribution*. The time has come to get to know this distribution curve better, for the characteristics of this distribution (and similar distributions) are of fundamental importance in much of statistical theory and its practical applications. This will become clear in the next four or five chapters, which all deal in some way with the *statistical significance* of experimental results.* *The statistical*

---

*Statistical significance should not be confused with *importance*. For example, even though the difference between the mean scores of an experi-

significance of experimental findings is usually expressed in probability terms. And these probabilities are commonly estimated from the normal z distribution (or similar distributions on the basis of similar theory). *These probability estimates are based on areas under various parts of this normal curve.*

1. Areas under the z Distribution Curve

Figure 8 represents a particularly useful normal curve, *the z distribution curve, which by definition has a mean of zero ($\bar{z}$ or $\mu = 0$) and a standard deviation of one ($\sigma = 1$). The total area under this curve is also equal to 1.000.* Partial areas between any given set of limits on the baseline are expressed as decimal fractions of this total area. The limits indicated in Figure 8 are of special interest. The areas under the curve between these limits (and

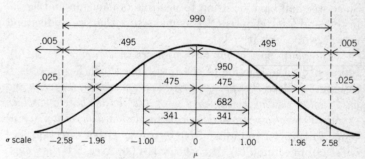

Figure 8. The z distribution curve, a normal distribution curve with $\sigma = 1.00$, area = 1.000, and mean, $\mu = 0$. Areas under the curve between certain important sets of limits, expressed in σ units, are given in decimal fractions of the whole area. Areas under the curve *beyond* some of these limits, in *the tails* of the distribution, are also given.

---

mental and a control group may be shown to be statistically significant, the difference may be so small as to be unimportant. On the other hand, such a difference may *appear* to be important and yet turn out to be not statistically significant. In this case it would be neither significant nor important.

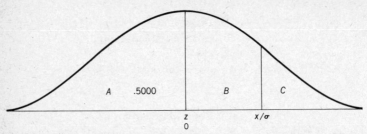

Figure 9. Total area under the z curve = 1.000. Area A = .5000. Area B + Area C (the upper tail) = .5000. Entries in columns (1) to (3) of Table 10 are for Area B, which always has the mean ($\bar{z} = 0$) as one of its limits. Column (4) of Table 10 lists selected values in the tail of the z distribution, Area C.

also the areas beyond these limits) are obtained from Table 10. Since we shall have occasion to use this table again and again, in this and later chapters, let us make sure that we understand how to work with it.

The main part of the table appears in columns (1) to (3), representing the fraction of the total area which lies in Area B of Figure 9. *Area B is always limited on one side by the mean and on the other side by the values of z (or x/σ) on the baseline, which represent deviations from the mean in standard deviation (or σ) units.* In theory z can range from 0 to ∞ in either direction; but we see from column (3) that when z is 4.00, Area B is very close indeed to one-half of the total area, which is its maximum value. It differs from 1.00000/2 = .50000 by only .00003. The left half of the normal curve, represented by Area A, is symmetrical with the right half, so we can get all the information we need about areas under the curve from Table 10.

Area C of Figure 9 is called the upper tail of the distribution, the area beyond the upper limit of Area B. Selected values of Area C are given in column (4). These and other values for a single tail of the distribution are obtained for any particular z score by subtracting the corresponding B Area from .500.

Now of course *z can be negative as well as positive.* If z is negative, Areas B and C would be below the mean and Area A above it. *But we use Table 10 in just the same way as described*

TABLE 10

The Distribution of z
Areas under the Normal Probability Curve
Corresponding to Distances on the Baseline
between the Mean and x/σ. Partial Areas Are Decimal Fractions
of the Total Area, Which = 1.000

(1)		(2)		(3)		(4)	
$z: x/\sigma$	Area B	x/σ	Area B	x/σ	Area B	x/σ	Area C[a]
0.00	.000	0.50	.192	1.75	.460	0.00	.500
0.02	.008	*0.525*	*.200*	1.80	.464	0.10	.460
0.04	.016	0.60	.226	1.85	.468	0.20	.421
0.06	.024	0.65	.242	1.90	.471	0.25	.401
0.08	.032	*0.675*	*.250*	*1.96*	*.475*	0.30	.382
0.10	.040	0.75	.273	2.00	.477	*0.385*	*.350*
0.12	.048	0.80	.288	2.05	.480	*0.525*	*.300*
0.14	.056	*0.84*	*.300*	2.10	.482	0.60	.274
0.16	.064	0.90	.316	2.15	.484	*0.675*	*.250*
0.18	.071	0.95	.329	2.20	.486	*0.84*	*.200*
0.20	.079	*1.00*	*.341*	2.25	.488	*1.00*	*.159*
0.22	.087	*1.036*	*.350*	2.30	.489	*1.036*	*.150*
0.24	.095	1.10	.364	*2.33*	*.490*	1.15	.125
0.26	.103	1.15	.375	2.40	.492	*1.28*	*.100*
0.28	.110	1.20	.385	2.45	.493	1.40	.081
0.30	.118	1.25	.394	2.50	.494	*1.645*	*.050*
0.32	.126	*1.28*	*.400*	2.55	.4946	1.80	.036
0.34	.133	1.35	.412	*2.58*	*.4951*	*1.96*	*.025*
0.36	.141	1.40	.419	2.65	.4960	*2.33*	*.010*
0.385	.150	1.45	.427	2.70	.4965	*2.58*	*.005*
0.40	.155	1.50	.433	2.81	.4975	2.81	.0025
0.42	.163	1.55	.439	3.09	.4990	3.09	.0010
0.44	.170	1.60	.445	3.30	.4995	3.30	.0005
0.46	.177	*1.645*	*.450*	3.70	.4999	3.70	.0001
0.48	.184	1.70	.455	4.00	.49997	4.00	.00003

[a] Entries in this column are *selected* values of *one* tail (the *upper tail*) of the z distribution. Other values for the C Area may be obtained by subtracting the values in the B Area (columns 1, 2, and 3) from .5000. The lower half of the normal curve is symmetrical with the upper half.

above, except that Area C would be the lower tail rather than the upper tail.

We can now check the areas represented by the arrows in Figure 8 against the entries in Table 10. The right-hand arrow of the inner (shortest) set represents the area limited by the mean ($\mu = 0$) and 1.00σ. Since this corresponds to Area B in Figure 9, we use the main part of Table 10. We find opposite $x/\sigma = 1.00$ in column (2) that the area involved is .341 of the total area. The symmetrical area on the left, limited by the mean and -1.00σ, would of course make up another .341 of the total area. So *between the limits $\mu - 1.00\sigma$ and $\mu + 1.00\sigma$ we have .682 (or approximately 68%) of the area under this normal curve.*

Similarly, for the next longer set of arrows we have on the right an area limited by the mean and 1.96σ. We find from column (3) of the table that this makes up .475 of the total area. The symmetrical area limited by the mean and -1.96σ makes up another .475 of the total area. Hence, *between the limits $\mu - 1.96\sigma$ and $\mu + 1.96\sigma$ we have the middle .950 (or 95%) of the area under this normal curve.*

And, finally, the longest set of arrows represents the symmetrical areas limited by $\mu - 2.58\sigma$ and $\mu + 2.58\sigma$. In column (3) of the table we find opposite the z score of 2.58 the entry .4951, which is usually rounded off to .495. So b*etween the limits $\mu - 2.58\sigma$ and $\mu + 2.58\sigma$ we have the middle .990 (or 99%) of the area under this normal curve.*

Figure 8 also has arrows which indicate areas lying outside the last two sets of limits, areas in the two tails of the distribution. As we shall see in later chapters, *we are just as much interested in the tails of the distribution as we are in the larger central areas.* The areas for each of the two symmetrical tails (represented by Area C in Figure 9) are listed in column (4) of Table 10. Opposite a z (or x/σ) of 1.96 we find .025, which means that .025 (or 2.5%) of the total area lies beyond a point 1.96σ units above the mean. And of course .025 (or 2.5%) of the area also lies below -1.96σ. So *for the two tails of this normal distribution beyond $\mu + 1.96\sigma$ and $\mu - 1.96\sigma$ we have a total of .05 (or 5%) of the whole area.* Similarly, using $z = 2.58$ in colunm (4), we can see that .005 of the whole area lies in each tail of the distribution. So for both tails of this normal distribution beyond $\mu \pm 2.58\sigma$ we have a total of .01 (or 1%) of the whole area.

We have just illustrated the use of Table 10 for finding two of the four kinds of area for which it is most frequently used. Before we go on to applications, let us have a short summary of how all four kinds of area are found.

Summary of Methods for Finding Areas under the z Distribution Curve from Table 10

Example 1. Area limited by the mean ($\mu = 0$) and any other given value of z (or x/σ) on the baseline. This corresponds to Area *B* in Figure 9. Read off areas directly from columns (1) to (3).

Example 2. Area beyond any given value of z, in one tail of the distribution. This corresponds to Area *C* in Figure 9. Read off directly in column (4), if listed, for the value of x/σ in question. Other values for one tail (Area *C*) may be obtained from the formula: Area $C = .500 -$ Area *B*.

Example 3. Entire area below any positive z score. Find Area *B* directly from columns (1) to (3) and then add .500.

Example 4. Entire area above any negative z score. Find the equivalent of Area *B* below the mean and add .500.

There are many useful applications of the characteristics of the normal curve of Table 10. These all involve the translation of this information into probability terms. The applications usually involve the assumption (which can often be checked) that an obtained set of scores is distributed in a reasonably good approximation to a normal curve; that is, that it is not badly skewed and certainly not bimodal.

If we can reasonably assume that a given set of scores, or any other variable, is normally distributed, we can then talk in probability terms, and *we can make predictions with various degrees of confidence.* We must first calculate the mean ($\bar{X}$) and the standard deviation (s). Then if we draw scores at random (by chance) from such a distribution,* we can say that the probability is .68 (or 68 in 100) that they will lie between the

---

*Scores must be drawn one at a time and replaced before the next draw.

limits $\bar{X} - 1.00s$ and $\bar{X} + 1.00s$. Or we can say that the probability is .95 (or 95 in 100) that the scores will lie between the limits $\bar{X} \pm 1.96s$. And the probability that the scores will lie between the limits $\bar{X} \pm 2.58s$ will be 99 in 100, which is usually expressed $p = .99$. This is one form of prediction. We shall see others later. We have already made use of the characteristics of a normal curve in Chapter 5 for "grading on the curve." In later chapters there will be extensive application of probability estimates based on Table 10 and similar distributions. In the next section we shall apply what we have just learned about the z distribution to estimating the probable position of the mean of a large population from the mean of a single sample. To do this we must first learn how to calculate the standard error of the mean, which is closely related to the variance of the mean.

2. The Variance and Standard Error of the Mean; Confidence Limits for the Mean

The *mean of a sample* set of scores drawn at random from a much larger *population* is often used, in conjunction with the standard error of the mean, to *estimate* the probable position of the *mean of the population*. In order to understand how such an estimate can be made we must make use of the z distribution again. Let us work with a concrete example. Suppose we wish to know the mean intelligence test score of all 12-year-old boys in a very large city school system. With a great expenditure of time and money, we could test every boy in this entire category, or population, totaling tens of thousands. Or we could work with samples chosen at random from this large population of 12-year-olds, chosen so as to avoid distortion due to such things as special opportunities or special handicaps in certain areas of the city. Extensive work with sampling has shown that, if we take a large number of samples of moderate size from a large population of this kind and calculate the mean for each of the samples, the means themselves tend to be distributed *around the mean of the population* in accordance with a normal probability curve. (This tendency becomes greater as the size of the samples increases;

it exists even though the distribution of the individual scores in the entire population is not exactly normal.) Since this is true, it is possible to treat the means of samples as individual scores, to work out a variance or a standard deviation for them, to interpret their variability in terms of probability, and to draw conclusions from them about the probable position of the mean of the population from which they were drawn in terms of *confidence limits*.

The variance of such a set of sample means may be *estimated* from a single random sample by *the variance of the mean*:

$$s_{\bar{X}}^2 = \frac{s^2}{n} \qquad (7.1)$$

where s^2 is the variance of the sample and n is the number of cases in the sample. By substituting in this formula the equivalent of s^2 from formula *4.2* (p. 38) we get a useful alternative expression for the variance of the mean in terms of *the sum of squares* of the sample:

$$s_{\bar{X}}^2 = \frac{\Sigma x^2}{n(n-1)} \qquad (7.2)$$

The standard deviation of the sample means is called *the standard error of the mean*. It is *estimated* from a single random sample by taking the square root of the variance of the mean. This gives us either

$$s_{\bar{X}} = \frac{s}{\sqrt{n}} \qquad (7.3)$$

or

$$s_{\bar{X}} = \sqrt{\frac{\Sigma x^2}{n(n-1)}} \qquad (7.4)$$

Although formula *7.4* looks more complicated, its use reduces the number of square roots required in the calculation of $s_{\bar{X}}$.

The standard error of the mean, $s_{\bar{X}}$, is used to determine the *confidence limits of the population mean*; that is, it is used to *estimate* how close the population mean, μ,* comes to the mean

---

*The Greek letter for *m, mu*, pronounced "mew" as out of cat.

of the sample, $\bar{X}$. To illustrate its use as a way of estimating the probable position of the population mean, μ, let us return to the intelligence test scores of our 12-year-old school boys. If, for a random sample, we find that $n = 200$, $\bar{X} = 80$, and $\Sigma x^2 = 159{,}200$, then by formula 7.4

$$s_{\bar{X}} = \sqrt{\frac{\Sigma x^2}{n(n-1)}} = \sqrt{\frac{159{,}200}{200 \times 199}} = \sqrt{4.00} = 2.0$$

We now use this value of $s_{\bar{X}}$, together with the mean of the sample ($\bar{X}$), to set up the *confidence limits* for the mean of the population (μ). For example, we can say that the true mean, μ, lies between the limits $\bar{X} - 1s_{\bar{X}}$ and $\bar{X} + 1s_{\bar{X}}$ (or, in this case, between 78 and 82) with a *limited* degree of confidence. Specifically, if we repeatedly made such statements on the basis of a great many other samples selected at random from the same population, we should expect to be right only about 68 times out of 100 (see p. 69). But we should expect to be right about 95% of the time if we said that the true mean lies between the limits $\bar{X} - 1.96s_{\bar{X}}$ and $\bar{X} + 1.96s_{\bar{X}}$. (In this case, between $80 - 1.96 \times 2.0$ and $80 + 1.96 \times 2.0$, or between 76.1 and 83.9.) And we can say with great confidence that μ lies between $\bar{X} - 2.58s_{\bar{X}}$ and $\bar{X} + 2.58s_{\bar{X}}$. (In this case, between the limits $80 \pm 2.58 \times 2.0$, or between 74.8 and 85.2.) Specifically, if we repeatedly made statements of this last type, we should expect to be right *in the long run* about 99% of the time.

In using these confidence limits for the true mean (μ), we are not suggesting that μ varies around the sample mean ($\bar{X}$). It doesn't, of course; it doesn't vary at all. It is a fixed parameter of the population, around which the sample means vary. The use of the familiar probability limits above is based on the theory that if the sample means lie within certain limits of μ with the specified probabilities, then *in the long run* μ will be no further away from any given sample mean than is indicated by the same limits for any given probability.

In setting up the confidence limits for the population mean above we were working with a moderately large sample for which n was 200. This was large enough to justify the use of the z distribution curve in estimating the probable position of μ

at the 95% and 99% levels of confidence. In the next chapter, which deals with small samples (usually 30 or less), we are going to base our probability estimates on the characteristics of *the t distribution* which, though very similar to this normal distribution when the samples are large, has a flatter shape for small samples. The t distribution should be used instead of the z distribution for setting up confidence limits for the mean when the samples are small.

Exercises

DATA: *Means and Standard Deviations for Three Achievement Tests*

English: $\bar{X} = 60$, $s = 10$
History: $\bar{X} = 66$, $s = 8$
Mathematics: $\bar{X} = 70$, $s = 20$

Note: In all exercises assume that the test scores are normally distributed.

1. For each of the tests find the score limits for the middle 68% of the scores.

2. For the English test find the score limits for: (a) the upper 2.5%, (b) the upper 0.5%, and (c) the middle 95% of the scores.

3. For the history test find the score limits for: (a) the upper 10%, (b) the lower 5%, and (c) the middle 50% of the scores.

4. For the English test what percent of the scores lie: (a) on or above 70, (b) on or above 50, (c) on or below 79.6, and (d) on or above 36.7? (Hint: first convert scores into z scores.)

5. For the history test what percent of the scores lie: (a) on or below 58, (b) on or below 74, (c) on or above 76.24, and (d) on or above 55.76? (Hint: first convert scores into z scores.)

6. What is the probability that a score taken at random from the mathematics test will fall on or above: (a) 90, (b) 109.2, and (c) 121.6? (First convert to z scores.)

7. What is the probability that a score taken at random from the English test will fall on or beyond the limits: (a) 47.2 and 72.8, (b) 43.55 and 76.45, (c) 36.7 and 83.3, and (d) 27 and 93?

8. The data for the English test are based on a sample of 100 scores drawn at random from a larger population. Find: (a) the standard error of the mean, $s_{\bar{x}}$, (b) the 95% confidence limits for the true mean, μ, and (c) the 99% confidence limits for μ.

9. The mathematics test data are based on a random sample of 100 scores. Find: (a) the 95% confidence limits for μ, (b) the 99% confidence limits for μ, and (c) the 90% confidence limits for μ.

8
The Significance of a Difference between the Means of Small Samples, the t Test

In the last chapter we became thoroughly familiar with the characteristics of the z distribution curve. We learned that areas under this curve could be interpreted in probability terms. And the probability concept was applied to confidence limits in estimating the position of the population mean from the mean of a *large random sample*. In this chapter we shall apply many of the same ideas to *testing the significance of a difference between two means*, one of the major devices for evaluating experimental and test results. Since we shall now be working with *small samples*, however, we shall base our probability estimates on a set of curves which are similar to, but not identical with, the z distribution curve. We shall make use of the t distribution. We shall use the t distribution, not because we are going to work with two samples at a time rather than only one as in Chapter 7, but because our samples are small, and the t distribution is based on small sample estimates of population variance. The general principles involved in this chapter also apply when the z distribution is used with large samples. In fact, for large samples (when n is, say, 200 or more) the values of z and t are almost the same for the probability values in common use.

1. The Significance of a Difference between the Means of Two Small Independent Samples; the *t* Test

One of the most important applications of the statistical tools so far developed is testing for significance the difference between two means. Indeed, in most research work some such test is crucial. Let us work with a concrete example. On the final examination in a certain course taken by a fairly large number of both men and women, the mean score for a random sample of 11 women ($\bar{X}_1$) is found to be 50.0 and the mean score for a random sample of 11 men ($\bar{X}_2$) is 44.0. *Apparently*, the women are superior to the men in this course, for $\bar{X}_1$ is greater than $\bar{X}_2$. But can we trust this superficial judgment? Is $\bar{X}_1$ *significantly* greater than $\bar{X}_2$? Or is it merely a chance effect? If we took other samples of men and women students from this course would the difference disappear, or perhaps appear in the opposite direction, with $\bar{X}_2$ greater than $\bar{X}_1$? The answers to these questions depend upon two general factors (in addition to freedom from distortion in drawing the samples): the size of the difference and the variability of the two samples. A difference might be significant if the scores were closely clustered about the mean in each sample, so that the distribution curves overlapped little, if any, as in Figure 10 (a). Or, on the other hand, a difference might not

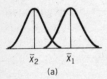
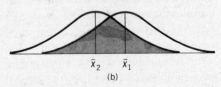

Figure 10.

be significant if the variability of the two samples was excessive, so that the distributions overlapped extensively, as in Figure 10 (b).

There are several methods in common use for testing the significance of a difference between two sample means. They usually involve *the t test*. This requires the calculation and evaluation of a statistic known as *t*, a quotient which takes into account both the extent of the difference between the means and the variability of the samples. For our present purposes *t* may be defined as the ratio of the difference between the sample means to the standard error of this difference; that is

$$t = \frac{\text{Diff.}}{s_{\text{Diff.}}} = \frac{\bar{X}_1 - \bar{X}_2}{s_{(\bar{X}_1 - \bar{X}_2)}} \qquad (8.1)$$

where $s_{(\bar{X}_1 - \bar{X}_2)}$ is the symbol for *the standard error of the difference* between the two means, $\bar{X}_1$ and $\bar{X}_2$. There are several formulas for it.

The simplest two of these *apply only when our samples are of the same size*; that is, when $n_1 = n_2 = n$. Then

$$s_{(\bar{X}_1 - \bar{X}_2)} = \sqrt{\frac{\Sigma x_1^2 + \Sigma x_2^2}{n(n-1)}} \qquad (8.2)$$

This may also be written

$$s_{(\bar{X}_1 - \bar{X}_2)} = \sqrt{\frac{2s^2}{n}}, \quad \text{where } s^2 = \left(\frac{\Sigma x_1^2 + \Sigma x_2^2}{n_1 + n_2 - 2}\right) \qquad (8.3)$$

Formula *8.2* is easier to work with, but formula *8.3* has certain advantages which will be discussed later (p. 83). In any case, *they yield the same numerical result*. Let us first work with formula *8.2*, which is easy to remember because the expression under the radical sign is obviously the sum of the variances of the means given by formula *7.2*. Continuing with our illustration, we shall calculate *t* for the examination results for the small samples of men and women ($n = 11$ for each) and see what we can infer about the larger groups, or populations, from which the samples were taken at random.

The calculation of t is relatively easy. (The evaluation is something else.) There are *four steps* in the calculation of *t* using *formulas 8.1 and 8.2*.

Step 1: Calculate *the two means.* Here they are given:

$$\bar{X}_1 = 50.0 \quad \text{and} \quad \bar{X}_2 = 44.0$$

Step 2: Calculate Σx_1^2 and Σx_2^2 by the most convenient method from Chapter 4, sections 3 and 4. Here these *sums of squares* are given:

$$\Sigma x_1^2 = 200 \quad \text{and} \quad \Sigma x_2^2 = 240$$

Step 3: Calculate *the standard error of the difference between the means.* Using formula *8.2*, we get for $n = 11$

$$s_{(\bar{X}_1-\bar{X}_2)} = \sqrt{\frac{\Sigma x_1^2 + \Sigma x_2^2}{n(n-1)}} = \sqrt{\frac{200+240}{11 \times 10}} = \sqrt{4.00} = 2.0$$

Step 4: Calculate *the t ratio.* Using formula *8.1*, we get

$$t = \frac{\bar{X}_1 - \bar{X}_2}{s_{(\bar{X}_1-\bar{X}_2)}} = \frac{50.0 - 44.0}{2.0} = 3.0$$

The Evaluation of t

Statisticians have shown that when t is based on the chance differences that occur between the means of many *pairs* of samples drawn at random *from the same population* of scores, t itself is distributed in a pattern that is a good approximation to a normal probability curve when large samples are used. *For small samples*, however, say 30 or less, the distribution of t differs from the normal pattern enough to require a special table of t on which to base our probability estimates. This will be introduced under the second of the *three steps in the evaluation of t*.

Step 1: We first *set up what is called a null hypothesis* (symbolized by H_o). In this case we temporarily assume that our two samples come *from the same population:* that is, we assume that there is *no difference between the true means* of the larger populations (of men and women) from which our sample sets of scores were drawn.* (This is symbolized by $H_o: \mu_1 = \mu_2$, which is read: "The null hypothesis is such that $\mu_1 = \mu_2$.") Setting up

---

*We also assume equal variances, but here we are primarily interested in the means.

this hypothesis that the samples come from the *same population* is a rather backward way of doing things, because usually we are interested in a possible *difference* between the means; we are looking for any evidence that H_o is probably false and that the samples come from *different populations*. We must, however, use this hypothesis of a common population source because the distribution of t as defined is based on the assumption that the *pairs* of samples are drawn from the *same* population.

Step 2: On the basis of our null hypothesis, we *determine the approximate probability that a value of t as large as the obtained value (or larger) could occur on the basis of chance* variations in the differences between pairs of samples drawn from the same population. To do this we must use Table 11 (next page.)

This is a *table of t*. All of the column headings (except the first) are probability (p) values. The rows give us the t values corresponding to the different p values. But *the t values differ from row to row depending on the size of the samples*. The sample size is indicated by the symbol $d.f.$, standing for *degrees of freedom*.* This is closely related to n, but is not quite equal to it. For a single sample $d.f. = n - 1$. For a pair of samples, as in our present problem, $d.f. = n_1 + n_2 - 2$.

Now let us see what the table tells us about the value of t calculated for our samples of 11 men and 11 women. Using $d.f. = 11 + 11 - 2 = 20$, and moving from left to right along this row, we find that our obtained value of t, 3.0, is a little larger than 2.85, the value listed under the heading $p = .01$. This means that the probability of getting a value of t as large as 3.0 from two random samples of this size from the same population is less than .01, which is indicated thus: $p < .01$.

Step 3: On the basis of the obtained probability value, we *draw a conclusion about the reasonableness of our null hypothesis*. If p is very small, our null hypothesis is rejected as improbable and our obtained difference between the sample means is regarded as significant (that is, we reject the hypothesis that the samples come from the same population and consider it probable that they represent two different populations). If, however, p is large, little doubt is cast on the hypothesis, and the obtained

---

*A suggestion as to the meaning of this concept is given in Chapter 15, section 3.

TABLE 11

Distribution of t[b]

d.f.	p[a] = .20	.10	.05	.02	.01	.002
1	3.08	6.31	12.7	31.8	63.6	318.3
2	1.89	2.92	4.30	6.97	9.93	22.3
3	1.64	2.35	3.18	4.54	5.84	10.2
4	1.53	2.13	2.78	3.75	4.60	7.17
5	1.48	2.02	2.57	3.37	4.03	5.89
6	1.44	1.94	2.45	3.14	3.71	5.21
7	1.42	1.90	2.37	3.00	3.50	4.79
8	1.40	1.86	2.31	2.90	3.36	4.50
9	1.38	1.83	2.26	2.82	3.25	4.30
10	1.37	1.81	2.23	2.76	3.17	4.14
11	1.36	1.80	2.20	2.72	3.11	4.03
12	1.36	1.78	2.18	2.68	3.06	3.93
13	1.35	1.77	2.16	2.65	3.01	3.85
14	1.35	1.76	2.15	2.62	2.98	3.79
15	1.34	1.75	2.13	2.60	2.95	3.73
16	1.34	1.75	2.12	2.58	2.92	3.69
17	1.33	1.74	2.11	2.57	2.90	3.65
18	1.33	1.73	2.10	2.55	2.88	3.61
19	1.33	1.73	2.09	2.54	2.86	3.58
20	1.33	1.73	2.09	2.53	2.85	3.55
21	1.32	1.72	2.08	2.52	2.83	3.53
22	1.32	1.72	2.07	2.51	2.82	3.51
23	1.32	1.71	2.07	2.50	2.81	3.49
24	1.32	1.71	2.06	2.49	2.80	3.47
25	1.32	1.71	2.06	2.49	2.79	3.45
26	1.32	1.71	2.06	2.48	2.78	3.44
27	1.31	1.70	2.05	2.47	2.77	3.42
28	1.31	1.70	2.05	2.47	2.76	3.41
29	1.31	1.70	2.05	2.46	2.76	3.40
30	1.31	1.70	2.04	2.46	2.75	3.39
40	1.30	1.68	2.02	2.42	2.70	3.31
60	1.30	1.67	2.00	2.39	2.66	3.23
120	1.29	1.66	1.98	2.36	2.62	3.16
∞	1.28	1.645	1.96	2.33	2.58	3.09

[a] The p values are for a *two-tailed test*. For a one-tailed test they should be divided by 2.

[b] This table is condensed from Table 12 of the *Biometrika Tables for*

difference, $\bar{X}_1 - \bar{X}_2$, cannot be regarded as significant (that is, a small value of t could well have occurred from chance variation in the samples drawn from the same population, as our hypothesis assumed).

In the case of our small samples of men and women, we conclude that the means of the larger populations of men and women taking the examination are not equal, $\mu_1 \neq \mu_2$. And it is a fairly safe bet that the mean of the women's population is larger than the mean of the men's population, $\mu_1 > \mu_2$, just as in the samples, $\bar{X}_1 > \bar{X}_2$. (See p. 92.) But such a conclusion is not inevitable. Under some circumstances we regard the null hypothesis as tenable and conclude that the difference between the means is not significant. Some criteria for making a decision about the fate of null hypotheses are discussed in the next paragraphs.

Criteria for the Rejection or Retention of a Null Hypothesis

A null hypothesis can never be rejected or retained with complete certainty. It can merely be shown to be highly probable, highly improbable, or something in between. *Two critical probabilities conventionally used* for testing hypotheses in the behavioral sciences are the $p = .01$ and the $p = .05$ values of t. If we reject the null hypothesis only when p is equal to, or less than, .01 ($p \leq .01$), we run very little risk of a false rejection. *In the long run* we would make a false rejection no more than about 1% of the time. This is a fairly rigorous criterion.

In fact, it may be too rigorous. If, in order to avoid a *false rejection of a true hypothesis* (called a *Type I error*), we insist on this criterion or something even more rigorous (like $p = .002$), we may fall into another kind of error. We may *fail to reject a false hypothesis* (which is called a *Type II error*). For example, if in our illustration above, the larger population of women was definitely better on the examination than the larger population of men ($\mu_1 > \mu_2$), and if we had used a more rigorous criterion

---

Statisticians, Vol. 1 (ed. 1), edited by E. S. Pearson and H. O. Hartley. Reproduced here with the kind permission of E. S. Pearson and the trustees of the Biometrika Trust. All entries are reduced to three significant figures, beyond which their use with most estimates becomes a little silly.

than we did use (say, $p = .002$), we would have failed to reject our null hypothesis (H_o: $\mu_1 = \mu_2$) when it was in fact false. (Because our obtained value of t was 3.0 and this was less than the 3.55 required for the rejection of H_o at the $p = .002$ level.) *We are less likely to make this second kind of error (Type II) if we use the $p = .05$ decision point.* With this criterion we *reject* the null hypothesis if p is equal to, or less than, .05 ($p \leq .05$), and we *regard the hypothesis as tenable* if p is greater than .05 ($p > .05$). This is *usually a good compromise between the risks of the two kinds of error. But a sound choice of appropriate criteria depends upon the degree of risk involved* in the false retention or rejection of the hypotheses in question. If, for example, a false decision involves the risk of losing life or limb, fortune or face, more rigorous criteria should be used. If the risk involved is slight, less rigorous criteria will suffice. The problem, however, is that *if the probability of a Type I error is reduced, the probability of a Type II error is increased*. A more detailed discussion of these matters will be found under *Decision Making* in Chapter 10.

It is interesting to note that the t test of significance does not require a completely normal distribution of scores in the parent populations; for the distribution of t itself tends to become normal if the samples are not too small (say, not under about 25), and *always provided that they are random samples*. The parent populations may, for example, be rather badly skewed; but if they are skewed in the same direction and degree, the test is still valid. For this and other reasons the t test is called *a "robust" test* of significance. Though deviations from the normal form in the original populations are less disturbing to the t test if the samples are of the same size, as assumed in the formulas used in this section, the samples do not *have* to be of equal size. A procedure for dealing with small samples of unequal size follows.

2. The Calculation of t for Small Independent Samples of Unequal Size

Though formulas *8.2* and *8.3* for *the standard error of the difference between two means* can be used only when $n_1 = n_2 = n$, the following formula proposed by R. A. Fisher is appropriate when samples are not of the same size.*

$$s_{(\bar{X}_1-\bar{X}_2)} = \sqrt{s^2\left(\frac{n_1+n_2}{n_1 \times n_2}\right)} \qquad (8.4)$$

where, as in formula *8.3*,

$$s^2 = \left(\frac{\Sigma x_1{}^2 + \Sigma x_2{}^2}{n_1+n_2-2}\right) \qquad (8.5)$$

It is useful to make the acquaintance of this quantity s^2 (which is, of course, the symbol for *variance*), for it will reappear in the chapters on Analysis of Variance in various forms. At the moment it is a necessary part of formula *8.4* and it has one other immediate advantage: the denominator, $n_1 + n_2 - 2$, gives us the proper *d.f.* to be used in the *table of t* when t is calculated from this formula combined with formula *8.1*.

Though formula *8.4* looks rather formidable, let us assume some values for the two n's and the two *sums of squares*, and see what happens. For $\Sigma x_1{}^2 = 160$, $\Sigma x_2{}^2 = 288$, $n_1 = 8$, and $n_2 = 14$, we get first from formula *8.5*

$$s^2 = \left(\frac{160+288}{8+14-2}\right) = \frac{448}{20}; \quad \text{and} \quad \left(\frac{n_1+n_2}{n_1 \times n_2}\right) = \frac{22}{112}$$

Substituting these values in formula *8.4*, we get

$$s_{(\bar{X}_1-\bar{X}_2)} = \sqrt{\frac{448 \times 22}{20 \times 112}} = \sqrt{4.40} = 2.10$$

Now all that we have to do to complete the calculation of t is to substitute this value in formula *8.1*, together with the value of $\bar{X}_1 - \bar{X}_2$, the difference between means we are testing for significance. And when we look up t in Table 11, we use *d.f.* $= n_1 + n_2 - 2$.

3. The Significance of a Difference between the Means of Two Correlated Samples

A correlation exists between two samples of scores when, for example, they represent: (1) the performance of *the same subjects*

---

*It is assumed that the population variances are *approximately* equal.

before and after the introduction of some experimental factor; or (2) the performance of *two groups matched*, subject by subject, with respect to some aptitude which affects the scores under consideration. A convenient method for testing the significance of a difference between the means of two such correlated samples is illustrated in Table 12. In spite of superficial differences in procedure and in the form of some of the formulas, the method of determining t is essentially the same as that of section 1. However, there is one important short cut: instead of calculating *two* means and *two* sums of squares and then the standard error of the difference between the means of the two samples, here we work with *a single mean only, the mean of the differences* in the performance of the subjects between condition X and condition Y. *Here our problem reduces to a test of whether the mean of the differences is significantly different from zero.* (The assumption that the true mean of the differences is zero is the null hypothesis in this case.)

In terms of the concrete data of this table our problem is to find out whether the "hearing scores" of the 12 subjects at an altitude of 16,900 feet are significantly different on the average from what they were at sea level; that is, to test whether the mean difference in scores ($\bar{D}$) is significantly different from zero. The *four steps for the calculation of t* are given below.

Step 1: We enter the subjects' scores for the two conditions tested, in parallel columns, and fill in the column of differences (D), *indicating any negative values* with minus signs; we then *calculate the mean of these differences* ($\bar{D}$) in the usual manner. (Though it is not necessary to calculate $\bar{X}$ and $\bar{Y}$, it is interesting to note that $\bar{D} = \bar{X} - \bar{Y}$; that is, the mean of the differences is equal to the difference between the means, as it should be, of course.)

Step 2: We next *calculate the sum of squares*, Σd^2, using the difference, D, as our variable (rather than the original scores, X and Y), by means of a familiar formula:

$$\Sigma d^2 = \Sigma D^2 - \frac{(\Sigma D)^2}{n} \tag{4.4}$$

Step 3: We then *calculate the standard error of the mean of*

TABLE 12

A Test of Significance for the Difference between the Means of Two Correlated Samples (a Check on the Apparent Loss in the Perception of Speech at an Altitude of 16,900 Feet)[a]

Subject	"Hearing Scores"		Difference $D = X - Y$	D^2
	Sea Level (X)	16,900 ft. (Y)		
1	21.0	9.0	12.0	144.00
2	37.0	22.0	15.0	225.00
3	5.0	9.0	−4.0	16.00
4	36.0	3.5	32.5	1056.25
5	18.0	1.0	17.0	289.00
6	36.0	3.5	32.5	1056.25
7	8.5	1.5	7.0	49.00
8	42.0	2.0	40.0	1600.00
9	15.0	0.5	14.5	210.25
10	37.0	33.5	3.5	12.25
11	7.5	3.0	4.5	20.25
12	32.0	11.5	20.5	420.25
$n = 12$	($\overline{X} = 24.6$)	($\overline{Y} = 8.3$)	$\Sigma D = 195.0$	$\Sigma D^2 = 5098.5$

(Step 1) $\quad \overline{D} = \dfrac{\Sigma D}{n} = \dfrac{195}{12} = 16.25$

(Step 2) $\quad \Sigma d^2 = \Sigma D^2 - \dfrac{(\Sigma D)^2}{n} = 5098.5 - \dfrac{(195.0)^2}{12} = 1929.7$

(Step 3) $\quad s_{\overline{D}} = \sqrt{\dfrac{\Sigma d^2}{n(n-1)}} = \sqrt{\dfrac{1929.7}{12 \times 11}} = \sqrt{14.62} = 3.82$

(Step 4) $\quad t = \dfrac{\overline{D}}{s_{\overline{D}}} = \dfrac{16.25}{3.82} = 4.26$

[a] Data slightly modified (to introduce a minus sign) from G. M. Smith and C. P. Seitz, "Speech Intelligibility under Various Degrees of Anoxia," *Journal of Applied Psychology*, 30, 1946, 182–191.

the differences ($s_{\overline{D}}$) by the familiar formula for the standard error of the mean:

$$s_{\overline{D}} = \sqrt{\dfrac{\Sigma d^2}{n(n-1)}} \qquad (8.6)$$

Step 4: Then we *calculate the t ratio,* in this case the ratio of the mean of the differences to the standard error of this mean, by the formula

$$t = \frac{\bar{D}}{s_{\bar{D}}} \qquad (8.7)$$

The evaluation of t involves the same considerations as those discussed in section 1. However, in using the table of t we must use $d.f. = n - 1$, where n is the number of *pairs* of scores, as in the calculations in Table 12. The *three steps* are as follows:

Step 1: *We set up a null hypothesis.* In this case $H_o: \bar{\Delta} = 0$, where $\bar{\Delta}$ (bar delta) stands for the mean of the differences *in the larger population* of "hearing scores," some scores from sea level and some from high altitude. Our H_o implies that it doesn't really matter whether the measurements were made at high altitude or at sea level; that *on the average,* counting both positive and negative values, differences between pairs drawn at random from the *same larger population* will be 0. *It implies that the obtained mean of the differences in our sample (16.25) was just a chance variation from 0.*

Step 2: From the table of t we *find the approximate probability that our obtained value of t could occur on the basis of chance.* Using $d.f. = n - 1$, or 11, we find that our obtained t of 4.26 > 4.03, the value listed under $p = .002$. In other words, $p < .002$ (or the chances are less than 1 in 500) that a t as large as 4.26 would occur on the basis of chance variations in the sample $\bar{D}$'s, if our null hypothesis is true.

Step 3: *We can, therefore, reject our null hypothesis with a very high degree of confidence;* and we conclude that $\bar{\Delta} \neq 0$. There is, in other words, a highly significant difference between the "hearing scores" of our subjects at sea level and at high altitude. And we may safely conclude that *the difference is in the obtained direction.* (See p. 92). The perception of speech is significantly worse at 16,900 feet.

Note: In the illustration above we used a technique for testing the significance of a difference between the means of two *correlated* samples. The samples happened to be small, but *the same procedure may be used with large correlated samples.*

However, as we shall see in the first section of the next chapter, when we work with large samples it is necessary to use only the last two rows of the table of t in order to determine the p values. And this simplifies matters.

Exercises

DATA: *Achievement Test Results for Random High School Samples*

(1) English, boys: $\bar{X}_1 = 60$, $\Sigma x_1^2 = 380$, $n_1 = 10$
(2) English, girls: $\bar{X}_2 = 68$, $\Sigma x_2^2 = 430$, $n_2 = 10$
(3) Mathematics, boys: $\bar{X}_1 = 65$, $\Sigma x_1^2 = 340$, $n_1 = 10$
(4) Mathematics, girls: $\bar{X}_2 = 60$, $\Sigma x_2^2 = 300$, $n_2 = 8$
(5) Scores in French, boys: 23, 21, 22, 25, 24, 21, 23, 25, 22, 24
(6) Scores in French, girls: 24, 23, 25, 27, 26, 24, 26, 27, 23, 25
(7) Hearing scores at 13,600 feet: 18, 14, 15, 14, 26, 22, 19, 12, 33, 22
(8) Hearing scores at 20,300 feet: 6, 8, 2, 5, 1, 2, 4, 0, 3, 14

Note: In Exercises 1–6 carry out the following steps in estimating the significance of a difference between the two means: (a) Calculate the standard error of the difference. (b) Calculate t. (c) State an appropriate null hypothesis. (d) Using the proper $d.f.$ in Table 11, determine the approximate probability that a value of t as large as the obtained value, or larger, could occur on the basis of chance variations in the samples. (e) Make an appropriate decision about the null hypothesis and draw the corresponding conclusion about the means of the populations from which the samples were taken, giving the rejection criterion used.

1. On the basis of the results in English (1 and 2) determine whether the mean for the girls is significantly greater than the mean for the boys.
2. Repeat Exercise 1 with $\bar{X}_2$ changed to 66. Compare results with Exercise 1.
3. Repeat Exercise 1 with Σx_2^2 changed to 260. Compare

results with Exercise 1. (f) In general, what is the effect of decreasing the variability of either sample on the significance of a difference between two means?

4. Using the results from the mathematics samples (3 and 4), find out if the boys are significantly superior to the girls.

5. Repeat Exercise 4 using $n_1 = 22$ and $n_2 = 20$. Does this modify the conclusion of Exercise 4? Explain.

6. Starting with original scores in the French test (5 and 6) for a sample of 10 boys and 10 girls, first calculate $\bar{X}_1$ (boys) and $\bar{X}_2$ (girls), Σx_1^2 and Σx_2^2. Then complete the test of significance as in the problems above.

7. Using the values of $\bar{X}_1$, $\bar{X}_2$, Σx_1^2, and Σx_2^2 obtained in Exercise 6, but assuming that $n_1 = 8$ and $n_2 = 12$, calculate t. Is the change in t enough to modify the conclusion in Exercise 6?

8. Using the 10 pairs of hearing scores (7 and 8), which are correlated (because each pair comes from the same subject), find out if the scores are significantly worse at the higher altitude, using the method of section 3.

9
A Significance Test for Large Samples; One-Tailed and Two-Tailed Tests

In the last chapter we worked with a variety of tests of significance designed primarily for small samples. We were introduced to two kinds of errors (Type I and Type II) that may be involved in making decisions about null hypotheses and in drawing conclusions about the significance of differences between sample means. In this chapter we shall deal with another rather simple t test, one for use with large samples. And we shall go a little further into the matter of making decisions on the basis of significance tests.

1. The Significance of a Difference between the Means of Two Large Independent Samples

This test of significance makes use of the same t ratio used in the last chapter with small independent, or uncorrelated, samples

$$t = \frac{\bar{X}_1 - \bar{X}_2}{s_{(\bar{X}_1 - \bar{X}_2)}} \qquad (8.1)$$

The only change is in the formula for *the standard error of the difference*, $s_{(\bar{X}_1-\bar{X}_2)}$. Instead of using the small sample formulas, here we use *the more general formula*

$$s_{(\bar{X}_1-\bar{X}_2)} = \sqrt{s_{\bar{X}_1}^2 + s_{\bar{X}_2}^2}$$

or $\quad s_{(\bar{X}_1-\bar{X}_2)} = \sqrt{\dfrac{\Sigma x_1^2}{n_1(n_1-1)} + \dfrac{\Sigma x_2^2}{n_2(n_2-1)}} \qquad (9.1)$

This allows for samples of unequal size. (If, however, $n_1 = n_2$, this formula reduces to formula *8.2*.) It should be obvious that the expression under the radical sign in formula *9.1* is (as before) the sum of the variances of the means (given by formula *7.2*). The calculation of t on the basis of formula *9.1* involves nothing really new or at all difficult. Furthermore, *the evaluation of t*, once we have obtained the value of p appropriate to our null hypothesis, involves the same principles as those used with small samples in section 1 of the preceding chapter.

In the case of two samples with a *combined n* of 150 or more our p values are easily determined, because for samples of this size the distribution of t closely approximates the normal distribution (Table 10, p. 68). In fact, for $d.f. = 150$, t values (in the range of probabilities of common interest) do not differ from the z values of the normal curve by more than about 1%. And when $d.f. = \infty$ (infinity) the distributions are *identical*. We can, therefore, determine our p values either from Table 10 (with a little juggling) or from the last row ($d.f. = \infty$) of Table 11, the table of t. *The most convenient procedure is to work with the last two rows of Table 11 using the following rule of thumb:* (1) when $d.f.$ ($n_1 + n_2 - 2$) is between 120 and 150, use the t values in the 120 row to obtain approximate p value; and (2) when $d.f.$ is 150 or more, use the t values in the last row (∞).

2. One-Tailed and Two-Tailed Significance Tests

In all our discussion of significance testing so far *we have evaluated t on the basis of a two-tailed test,* though we haven't called it that.

What does this mean? It means that the p values listed in our table of t, Table 11, are the probabilities of t deviating from its mean value of 0 by the indicated amount, or more, *in either direction*. It means that when we rejected a null hypothesis (that $\mu_1 = \mu_2$) at some indicated p level, we were counting the *combined probabilities for both positive and negative values of t*. For example, Table 11 tells us that for 30 d.f. the probability of an *absolute* value of t of 2.04 or more is .05. This p value is the sum of the probabilities in the two tails, the shaded areas, of Figure 11: $p = .025 + .025 = .05$. If t falls in *either* of the shaded areas, the null hypothesis can be rejected at the $p < .05$ level, if we decide on this criterion for a particular problem. Similarly, Table 11 tells us that for 30 d.f. the probability of an *absolute* value of t of 2.75 or more is .01. This is *the sum of* the probabilities in the two tails again, but the tails are smaller in this case: $p = .005 + .005 = .01$. This provides a more rigorous criterion of rejection for H_o, if this is what we decide we want.

If we are working *with large samples, with d.f. = 150 or more, we should use the normal curve to determine our probabilities*, or the bottom row of Table 11 where d.f. = ∞. This tells us that the probability of an *absolute* value of t, or z, (*in either direction*) of 1.96, or more, is .05; and that for an *absolute* value of 2.58, or more, $p = .01$. (These values should be familiar from our earlier discussion of areas under the normal curve, p. 68.)

Now *when do we use the two-tailed test?* It is used *in most scientific work* where the main interest is simply *in finding out what happens* when we introduce some experimental factor.

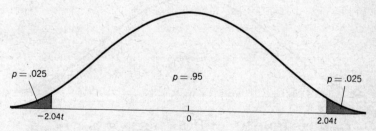

Figure 11. Two-tailed test of $H_o: \mu_1 = \mu_2$, or $H_o: \mu_1 - \mu_2 = 0$, at the $p = .05$ level. The null hypothesis is rejected if t falls in either shaded area. The values of t are for 30 d.f. only.

92 A Significance Test for Large Samples

Although we may have a hunch as to which way the wind will blow (as in the high altitude experiment in section 3 of the last chapter), we are interested (or should be) in any significant difference between the means of our experimental and control groups *regardless of the direction of the difference*. In such a case the proper null hypothesis is the familiar $H_o: \mu_1 = \mu_2$. If we reject this on the basis of some chosen criterion, we conclude that either $\mu_1 > \mu_2$ or *vice versa*. It is a fairly safe bet, however, that if a difference between the population means does exist, it is in the same direction as the obtained difference between the sample means, $\bar{X}_1$ and $\bar{X}_2$, if we have used either the $p = .05$ or $p = .01$ criterion of rejection.

The one-tailed or directional test is most useful in practical problems (and some theoretical ones) in which we are concerned with a difference between means *in one direction only*. Suppose, for example, that we have in our laboratory a number of expensive timing devices made by company 2. A salesman from company 1 tries to get us to change to his make of timer. But before we replace ours with company 1 timers we must be convinced that the rival product is definitely *better* in some way (more accurate, more durable, etc.), *not just as good, and certainly not worse*. The salesman shows us some figures based on an impartial accuracy test with 16 timers chosen at random from each make of timer. The mean accuracy of his company 1 timers is somewhat better than the mean accuracy of the company 2 sample. Since the durability and the price are about the same,

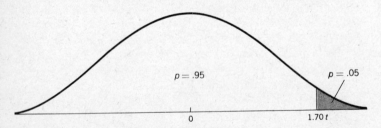

Figure 12. One-tailed test of $H_o: \mu_1 \leq \mu_2$, or $\mu_1 - \mu_2 \leq 0$, at the $p = .05$ level. The null hypothesis is rejected if t falls in the shaded area. The value of t is for 30 d.f. only.

he thinks he has a strong case for the replacement of our company 2 timers. But he is not a statistician! We proceed to get some information about the *variances* of the accuracy test samples, in addition to what he has shown us about the difference between the means: $\bar{X}_1 > \bar{X}_2$. And we calculate a t score, which comes out 1.75.

We then evaluate this t by means of a one-tailed test. This is illustrated in Figure 12. Our null hypothesis is that the general average accuracy of company 2 timers (μ_2) is equal to, or better than, the general average accuracy of the timers made by the rival company 1 (μ_1); that is, $H_o: \mu_1 \leq \mu_2$. If we have sufficient evidence to reject this, we can conclude that $\mu_1 > \mu_2$; that is, that our company 2 timers are in general less accurate than those made by the company which the salesman represents. Now, since our t of $1.75 > 1.70$ and, therefore, falls in the rejection area, we *could* draw this conclusion (with a long run probability of error of less than .05). However, this conclusion is not a certainty and our present timers still have some life left in them. Since replacing them will involve a large expense, we may decide that the evidence of the superior accuracy of the company 1 timers (at the $p < .05$ level) is not good enough. We tell the salesman that if he can show that his timers are better at the $p < .01$ level, we will raise the money for them. At this point he leaves the office looking not only disappointed but very much confused.

There is one thing (at least) that may have confused a lot of other people too in this discussion. Where did the 1.70 value of t come from? Why is this the boundary for the $p = .05$ rejection area? We had previously found that $t = 2.04$ corresponded with $p = .05$ for 30 $d.f.$ and Table 11 has 1.70 listed, for 30 $d.f.$, under $p = .10$, not .05. The answer is that the p values in our table of t are for a two-tailed test. *For a one-tailed test all the p values in Table 11 should be divided by 2.*

In our timer illustration above, the rejection area for t was in the right or positive tail of the t distribution (Figure 12). It is perfectly possible, of course, to have the rejection area for t on the left or negative side as in Figure 13. This would apply when the null hypothesis was $\mu_1 \geq \mu_2$, or $\mu_1 - \mu_2 \geq 0$. (Though

94 A Significance Test for Large Samples

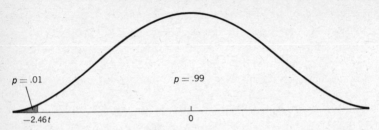

Figure 13. One-tailed test of H_o: $\mu_1 \geq \mu_2$, or $\mu_1 - \mu_2 \geq 0$, at the $p = .01$ level. The null hypothesis is rejected if t falls in the shaded area. The value of t is for 30 $d.f.$ only.

in our timer illustration the rejection area was in the positive tail, we can see from this figure, which involves a $p = .01$ rejection criterion, that our salesman in order to make the sale would have had to produce some very strong evidence of a superior product: $|t| > 2.46$, not just greater than 1.70.)

In deciding between a directional and a two-tailed test, we should consider something besides the nature of the problem under investigation, scientific or practical. *We should consider the relative power of the two tests to protect against Type II errors* (failure to reject false null hypotheses). Other things being equal, *the one-tailed test has somewhat more power than the two-tailed test, but this is mainly in one direction only.* Specifically, the greater protection against Type II errors occurs when *the true difference* between the population means ($\mu_1 - \mu_2$) is on the same side of zero as the rejection area for the null hypothesis (both positive or both negative.) On the other hand, if $\mu_1 - \mu_2 < 0$ (or negative) and the rejection area is far over on the positive side, there is little likelihood that H_o will be rejected and the correct conclusion be drawn. That is, there is little power (or protection) against a Type II error.

Though *the two-tailed test* has a little less power than the one-tailed test in one direction, it *has power against Type II errors in both directions.* That is, there is protection when $\mu_1 - \mu_2$ is either positive or negative. See Figure 14.

At this point, the business of protection against Type II errors may seem a little academic. Surely there must be more

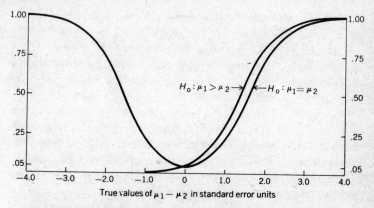

Figure 14. Comparison of the power curves for a one-tailed and a two-tailed test (when the maximum probability of a Type I error is set at .05). Power against Type II errors, indicated on the Y axis, varies greatly with the true values of $\mu_1 - \mu_2$, for either kind of test.

deadly sins. But when it comes to *decision making*, discussed more fully in the next chapter, we shall see that some control over both Type I and Type II errors, and the risks involved in each, can at times have important implications.

Exercises

DATA: *Various Comparisons of the Means of Two Large Samples*

A: $\bar{X}_1 = 50.0$, $\bar{X}_2 = 46.4$, $\Sigma x_1^2 = 9940$, $\Sigma x_2^2 = 5100$, $n_1 = 71$, $n_2 = 51$.

B: $\bar{X}_1 = 80.0$, $\bar{X}_2 = 84.7$, $\Sigma x_1^2 = 4970$, $\Sigma x_2^2 = 19{,}440$, $n_1 = 71$, $n_2 = 81$.

C: $\bar{X}_1 = 30.0$, $\bar{X}_2 = 20.0$; $\Sigma X_1 = 3000$, $\Sigma X_1^2 = 129{,}600$, $\Sigma X_2 = 1200$, $\Sigma X_2^2 = 41{,}700$; and $n_1 = 100$, $n_2 = 60$.

D (grouped and coded data): $\bar{X}_1 = 254.8$, $\bar{X}_2 = 250.0$; $\Sigma f x_1^2 = \Sigma x_1^2 = 19{,}440$, $n_1 = 81$; and $i_2 = 10$, $\Sigma f X_2' = 2000$, $\Sigma f X_2'^2 = 40{,}099$, $n_2 = 100$.

Note: For all exercises, when testing for significance, *use the same five steps* called for in the note above the exercises in Chapter 8, p. 87. But use the *more general* formula for the standard error of the difference (9.1).

1. If $\bar{X}_1$ and $\bar{X}_2$ in comparison A are the means of two experimental groups in a scientific experiment on reinforcement in learning, do the means differ significantly? Justify your answer and your choice of a one-tailed or two-tailed test.

2. If in a practical problem the results in comparison A are of interest only if $\bar{X}_1$ is significantly greater than $\bar{X}_2$, how many tails should be used in the significance test? Is $\bar{X}_1$ significantly greater than $\bar{X}_2$? Justify your answer. (Can you explain the apparent contradiction between the conclusions in this exercise and the last for the $p < .05$ rejection criterion?)

3. If comparison B involves the mean achievement scores of random samples from the junior and senior classes, can we reasonably conclude that the mean of one class differs significantly from that of the other? Explain your conclusion, giving the rejection criterion used, and your choice of a one-tailed or two-tailed test.

4. A prize has been offered by some tight-fisted alumnus to the Senior Class Fund if the seniors' achievement scores are significantly higher than those of the junior class, using the $p < .01$ rejection criterion. If $\bar{X}_2$ in comparison B represents a random sample of the seniors, will the prize be given? Explain choice of one-tailed or two-tailed test. Compare conclusion with that of Exercise 3. (Can you explain why the same t value satisfies a more rigorous rejection criterion in one case than in the other?)

5. $\bar{X}_1$ and $\bar{X}_2$ in comparison C represent the mean track and field scores of random samples of 100 20-year-old males and 60 20-year-old females. Is there a significant difference between the means? At what p level? What type of significance test should be used to establish such a difference?

6. What type of significance test should be used to see if $\bar{X}_1$ (males) is significantly greater than $\bar{X}_2$ (females)? If $\bar{X}_1$ is significantly greater, at what p level (comparison C)?

7. In comparison D the original data was grouped and coded. For Group 2 first find $\Sigma f x_2^2$, which is the equivalent

of $\Sigma x_2{}^2$ for ungrouped data, by formula 4.5. Then see if there is a significant difference between the means, using a two-tailed test. Give level of significance, if any.

8. If $\bar{X}_1$ and $\bar{X}_2$ are the mean scores on a composite test of physical skills for random samples from a given age group from Tribes A and B, at what p level is Tribe A superior to Tribe B (comparison D)?

10
Decision Making and the Power of Significance Tests[*]

In the last chapter, section 2, we have already had two examples of decision making, or *deciding what criteria to use in making decisions*. In one the decision was to use the more rigorous $p < .01$ rather than $p < .05$ as the criterion for rejecting the null hypothesis (about the two makes of laboratory timer). In the other the decision was between one-tailed and two-tailed tests. And part of this last decision had to do with the power of the tests to protect against Type II errors. Back in Chapter 8 (p. 81) a hint was given of the dilemma we find ourselves in when trying to avoid both Type I and Type II errors at the

---

[*]This chapter may be omitted, at the discretion of the instructor, without serious loss of continuity, but not without losing some important implications of significance tests.

same time. In this chapter we shall go a little further into ways of dealing with this dilemma.

1. Decision Making and Calculating Test Power

At the outset we must realize that *there is no fool-proof solution*. As with any dilemma, we may get hooked on either horn. The best we can do is to play the percentages. The problem is that if we decrease the probability of a Type I error, which is easy to do, we increase the probability of a Type II error, other things (such as sample size) being equal. In making a decision we have to consider the relative importance of the risks involved, if we have any way of evaluating the risks.

In discussing these matters it is helpful to review the definitions of these two kinds of error, to learn the definition of *power*, and to use the *conventional symbols* for the probabilities of each of these.

(1) A Type I error is the rejection of a true H_o. The probability of this kind of error is symbolized thus:

$$p \text{ (Type I error)} = \alpha \text{ (alpha)}$$

(2) A Type II error is the failure to reject a false H_o. The probability of this kind of error is symbolized thus:

$$p \text{ (Type II error)} = \beta \text{ (beta)}$$

(3) The *power* of a significance test is the probability of *avoiding* a Type II error, symbolized thus:

$$\text{power} = 1 - \beta$$

As has been said, it is very easy to reduce the probability of a Type I error. *We can make α as small as we like.* We simply decide to make the rejection areas for our null hypothesis (H_o) small, using as our cut-off point $p = .01$ or $p = .002$, or an even smaller p value. *But if we do this, our test usually loses power; that is, the probability of a Type II error (β) increases.* In failing to reject a null hypothesis at some more modest level (like $p = .05$ or even $p = .10$), *we may be missing some useful ex-*

perimental effect, some true difference between the population means of our experimental groups, $\mu_1 - \mu_2$.

Now in order to make a decision as to what rejection criterion to use for H_o, we have to have some way of estimating the effect of α on β and the power of the test $(1 - \beta)$. This is not a simple matter, for *the power of a test is dependent, not only on α, but also on the size of the samples (n), the variance of the parent populations (σ^2), and any true difference that may exist between the means of the parent populations,* $\mu_1 - \mu_2$. There are ways of determining or estimating n and σ^2 (see section 2). *The crux of the problem is the true difference, if any, between the population means, $\mu_1 - \mu_2$, which we don't know.* In fact, the whole purpose of the significance test is to find out the probability that such a difference exists at all. This looks like an incurable headache, but it isn't, as we shall see.

Let us examine the problem in three parts, illustrating it with a concrete (if somewhat fanciful) experimental case. Under Part I we shall simply apply the familiar two-tailed significance test to given experimental data, using a conventional criterion for the rejection of H_o, $\alpha = .05$. Under Part II we shall make certain assumptions about the true difference between the population means, $\mu_1 - \mu_2$, and then calculate the power of our test. And in a third part in the next section we shall see how the power of our test can be improved. Then we should be in a better position to decide on the criteria for making ultimate decisions. The ultimate decisions are those involving scientific truth or practical considerations of dollars, psychological adjustment, prestige, or convenience, etc.

In the illustrations that follow *we are going to make use of the distribution of z, the normal distribution, Table 10 (p. 67) in determining our probabilities*, even though n may be a little small, because this table is more complete than the table for the t distribution. *The same principles apply to small samples and the t distribution*, for which more complete tables are available if necessary.

Part I: The Conventional Approach

Let us suppose that a food chemist has been working for a long time on the development of a diet ingredient which he hopes will safely control excess weight in women. In a pilot study with laboratory rats he has obtained some promising results. His next step, before rushing into the expense and possible risk of marketing his product for humans, is to repeat his experiment with female house cats. He tests 30 experimental subjects before and after a six-week period in which "fatex" is included in their standardized diet. He also tests 30 control subjects for the same period. Their diet and exercise routine are the same as those of the experimental group, except that they are not supplied with "fatex." Using a composite index which includes weight changes, activity levels, etc., he gets the following results, a higher score indicating a more desirable response.

Mean score of the experimental group, $\bar{X}_1 = 13.30$
Mean score of the control group, $\bar{X}_2 = 10.00$

The variances of the two groups are approximately equal and yield an estimate of the population variance of $\sigma^2 = 60$. Since $n_1 = n_2 = n = 30$, the standard error of the difference between the means, as estimated by formula *8.3* is

$$\sigma_{(\bar{X}_1 - \bar{X}_2)} = \sqrt{\frac{2\sigma^2}{n}} = \sqrt{\frac{2 \times 60}{30}} = 2.00$$

When we use the z distribution to determine probabilities, rather than the distribution of t, the test statistic is

$$z = \frac{\bar{X}_1 - \bar{X}_2}{\sigma_{(\bar{X}_1 - \bar{X}_2)}} \qquad (10.1)$$

In this case

$$z = \frac{13.3 - 10.0}{2.00} = 1.65$$

The critical value of z for the conventional $\alpha = .05$ criterion of rejection for a two-tailed test is 1.96. (From Table 10, column 4, we see that a z score of 1.96 cuts off .025 of the total area *in one tail;* and, therefore, for both tails we have $\alpha = .05$.) Now, since the obtained z score of $1.65 < 1.96$, the chemist does not reject but retains the null hypothesis that $\mu_1 = \mu_2$. And he concludes that the difference between the mean responses of the experimental and control cats is not significant. The diet pill that he has labored so long to develop does not appear to produce a significant result. This is a bitter pill for him to swallow.

But before he abandons his pet project he calls in a statistician. The statistician at once checks on the possibility of a Type II error, the chemist's failure to reject H_o and discover a probable difference between the means of the parent populations, of which the experimental and control cats were a random (or stray) sample. He points out to the chemist that if he had used $\alpha = .10$ instead of .05, his obtained z score would have been just large enough to justify the rejection of his H_o, 1.65 $\geq$ 1.645 (see Table 10, column 4), and he would have concluded that there was a probable difference between the parent feline populations after all. The chemist was already aware of this, but he did not want to risk a .10 probability of a false claim for the virtues of "fatex." Then the statistician wants to know if the power of the chemist's test, even though based on $\alpha = .05$, is what it should have been. He makes use of the principles and procedure brought out in Part II.

Part II: Calculating the Power of a Significance Test

The power of a significance test cannot be determined in general; it can only be determined for a particular choice of α, a particular standard error of the difference, and for a particular true difference between the population means, $\mu_1 - \mu_2$. Now since there is no way of knowing $\mu_1 - \mu_2$, we seem to be stopped before we start. But we get around this road block by *assuming* various values of $\mu_1 - \mu_2$ and then calculating the power for each assumed value. The results can be useful in the process of decision making.

Continuing with our case of the chemist and the cats, we shall calculate the power of his test for two assumed values of $\mu_1 - \mu_2$ to illustrate the procedure and to see some of the implications for his experiment. He has already found that with $n_1 = n_2 = n = 30$ and $\sigma^2 = 60$, $\sigma_{(\bar{X}_1 - \bar{X}_2)} = 2.00$. He has decided to set $\alpha = .05$. This is the clear risk he takes of a Type I error, the probability of rejecting $H_o: \mu_1 - \mu_2 = 0$, which may be true.

Decision Making and Calculating Test Power

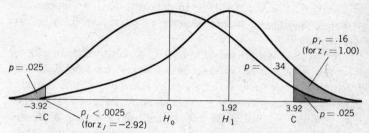

Figure 15. Rejection areas in two sampling distributions of $\bar{X}_1 - \bar{X}_2$, when $\sigma_{(\bar{X}_1 - \bar{X}_2)} = 2.00$ and $\alpha = .05$. The curve on the right represents the distribution of $\bar{X}_1 - \bar{X}_2$ when the *true* difference between the means, $\mu_1 - \mu_2 = 1.92$. The curve on the left is the distribution on the basis of $H_o: \mu_1 - \mu_2 = 0$. The rejection areas in *both* curves are limited by *the common cut-off points, C and $-$C, which are based on the assumption that H_o is true.* For $\alpha = .05$, C is 1.96 $\sigma_{(\bar{X}_1-\bar{X}_2)}$ above the zero mean of the H_o curve. In raw score units this is $1.96 \times 2.00 = 3.92$. But since the mean of the H_1 curve is 1.92, C is only $3.92 - 1.92 = 2.00$ *raw score units* above this mean. *In standard error units* C is $2.00/2.00 = 1.00$ above the mean of H_1. Thus $z_r = 1.00$ (see text). z_l is determined in a similar manner by subtracting 1.92 from $-$C.

But it is only by rejecting H_o, which may also be false, that he can avoid a Type II error, the failure to reject a false H_o. Power, the probability of avoiding a Type II error, is thus based in part upon the probability of rejecting H_o. Figure 15 illustrates how this works out for $\alpha = .05$, $\sigma_{(\bar{X}_1-\bar{X}_2)} = 2.00$, and IF we take the true $\mu_1 - \mu_2$ as $= 1.92$. If this is the case, sample mean differences, $\bar{X}_1 - \bar{X}_2$, will in reality be normally distributed about the population mean difference of $\mu_1 - \mu_2 = 1.92$ and not about $\mu_1 - \mu_2 = 0$, as H_o assumes. The center of the true distribution (H_1), which represents a real population difference, is 1.92 raw score units above the center of the H_o distribution, which assumes no difference in the populations from which the experimental and control samples were drawn.

Since power is based on the probability of rejecting H_o, the trick in calculating power is to find the rejection areas in the true distribution of $\bar{X}_1 - \bar{X}_2$ which are determined by the rejection areas (or their cut-off points) in the H_o distribution. This is done in *four easy steps*.

Step 1. Find the cut-off points C and $-C$ (Figure 15) for the rejection areas required by α *in terms of raw score deviations from 0, the center of the $\bar{X}_1 - \bar{X}_2$ distribution based on H_o.* In this case $\sigma_{(\bar{X}_1-\bar{X}_2)} = 2.00$ and $\alpha = .05$. Therefore, for a two-tailed test, $C = 1.96\ \sigma_{(\bar{X}_1-\bar{X}_2)} = 1.96 \times 2.00 = 3.92$, and $-C = -3.92$. (See Table 10, column 4.)

Step 2. Determine the z score equivalents of C and $-C$ for *the true distribution of $\bar{X}_1 - \bar{X}_2$ based on H_1.* We shall call these cut-off points for the right and left tails z_r and z_l. The formulas are obvious:

$$z_r = \frac{C - (\mu_1 - \mu_2)}{\sigma_{(\bar{X}_1-\bar{X}_2)}}$$

and

$$z_l = \frac{-C - (\mu_1 - \mu_2)}{\sigma_{(\bar{X}_1-\bar{X}_2)}}$$

where $\mu_1 - \mu_2$ is the true difference between the population means. In this case

$$z_r = \frac{3.92 - 1.92}{2.00} = \frac{2.00}{2.00} = 1.00$$

and

$$z_l = \frac{-3.92 - 1.92}{2.00} = \frac{-5.84}{2.00} = -2.92$$

Step 3. From the table of z (Table 10) *determine the size of the rejection areas in the tails beyond z_r and z_l.* We shall use the symbol p_r to indicate the probability of $\bar{X}_1 - \bar{X}_2$ being equal to or greater than z_r; that is p_r for $p(\bar{X}_1 - \bar{X}_2 \geq z_r)$. And we shall use the symbol p_l for $p(\bar{X}_1 - \bar{X}_2 \leq z_l)$. In this case we find from Table 10, column 4, that for $z_r = 1.00$, $p_r = .16$; and for $z_l = -2.92$, $p_l < .0025$, which is negligible. We notice that when the true $\mu_1 - \mu_2$ is positive, the z_r cut-off point is closer to the mean of the true distribution than z_l is, much closer in this case. Hence, p_r is much larger than p_l.

Step 4. The power is the sum of p_r and p_l. In this case

$$\text{power} = p_r + p_l = .16 + .00 = .16, \text{ approx.}$$

The power of the chemist's test, if the true $\mu_1 - \mu_2$ had been 1.92 (which is close to 1 standard error) was painfully low. His chances of making a Type II error were, therefore, painfully large: $\beta = 1 - \text{power} = 1 - .16 = .84$.

The fault in the chemist's program is perhaps not his wonder pill, after all, but in his significance test or in his experimental procedure. But before we suggest ways of improving these, *let us calculate the power of his test for a larger true $\mu_1 - \mu_2$, say 3.30*. This is the same size as the difference between the sample means, $\bar{X}_1 - \bar{X}_2$, which he actually found in his experiment.

Step 1. For the same $\alpha = .05$ and $\sigma_{(\bar{X}_1-\bar{X}_2)} = 2.00$, C remains 3.92 and $-C$ remains -3.92.

Step 2.

$$z_r = \frac{C - (\mu_1 - \mu_2)}{\sigma_{(\bar{X}_1-\bar{X}_2)}} = \frac{3.92 - 3.30}{2.00} = .31$$

and

$$z_l = \frac{-C - (\mu_1 - \mu_2)}{\sigma_{(\bar{X}_1-\bar{X}_2)}} = \frac{-3.92 - 3.30}{2.00} = -3.61$$

Step 3. From Table 10, column 1, $p_r = .50 - .12 = .38$, and from Table 10, column 4, $p_l < .0005$, which is negligible.

Step 4. The power, $p_r + p_l = .38$.

The power has improved somewhat for the larger value of the assumed true difference between means, $\mu_1 - \mu_2$. But the chances of a Type II error are still much too large: $\beta = 1 - \text{power} = 1.00 - .38 = .62$.

When the assumed true $\mu_1 - \mu_2$ is 4.00, the power works out to be .52 and $\beta = .48$. This means that even when the true $\mu_1 - \mu_2$ is two standard errors above 0, the value called for by the null hypothesis, the chances are just about even (48 to 52) that the chemist will fail to reject H_o and will conclude that his pill is worthless. Though the power of test improves as the assumed true value of $\mu_1 - \mu_2$ is increased to a rather large value (perhaps unrealistically large), he is still running too much of a risk of being let down by his present significance test. Now what can be done to improve it? Methods for increasing test power are taken up in the next section.

2. How To Increase the Power of a Significance Test

There are *three ways* to increase test power and thus cut down on the probability of a Type II error: (1) we can increase α, the probability of a Type I error; (2) we can increase the size of the samples, n; and (3) we can decrease the estimated population variance, σ^2.

Method I: Increasing α

We have already seen in the section above that the food chemist could have concluded that his weight control pill was effective with his female cats if he had used $\alpha = .10$, rather than $\alpha = .05$, as his criterion for rejecting the null hypothesis (p. 102). He would have run less risk of a Type II error, but more risk of a Type I error. He decided that the cost of a Type I error was greater, so he kept $\alpha = .05$. He was probably wise not to take a greater risk of law suits for false claims for "fatex" or not to risk the cost of promoting this new product without fairly substantial evidence of its value. But there are situations in which an experimenter may want to take a larger risk of a Type I error, and may even set $\alpha = .20$, in order to increase the power of his test.

Let us see, for example, what the power becomes in our chemist's significance test if all the conditions are kept the same as in the second example in Part II above except that α is set at .20, rather than at .05: $\sigma_{(\bar{X}_1-\bar{X}_2)} = 2.00$ (based on $n = 30$ and $\sigma^2 = 60$) and $\mu_1 - \mu_2$ is assumed to be 3.30. Then

$$C = 2.00 \times 1.28 = 2.56 \text{ (See Table 10, column 4)}$$

$$z_r = \frac{2.56 - 3.30}{2.00} = -.37$$

$p_r = .50 + .14$ approx. (Table 10, column 1) $= .64$

z_l is so far out on the left that

p_l is negligible

Therefore, power $= .64$ (and $\beta = 1 - .64 = .36$).

In the earlier example when α was set at .05, the power was only .38. So by setting $\alpha = .20$ the power has been increased by nearly 70%. The decision to set α at such a large value is of course up to the experimenter, after weighing the risks of the two kinds of error. But *it is possible to increase the power even more and to keep α small at the same time, by increasing the size of the samples.* This is explained below.

Method II: Estimating Sample Size for Desired α, Maximum β, and a Minimum $|\mu_1 - \mu_2|$ that Will Be Worthwhile

It is possible *to estimate* the size of the samples required for a specified risk of a false rejection of H_o (α), while at the same time not allowing the risk of failure to reject H_o which may be false (β) to exceed a specified maximum. As with any calculation of test power, some assumption about the true population means, μ_1 and μ_2, must be made. In this case, if we include in our calculation some specific minimum *absolute* value of $\mu_1 - \mu_2$ that will be either of practical or theoretical importance, we may estimate the required size of our two samples from the following formula:

$$n = \frac{2\sigma^2}{\delta^2}(z_\alpha + z_\beta)^2 \qquad (10.2)$$

where σ^2 is the estimated population variance and δ is the minimum $|\mu_1 - \mu_2|$ of interest to the experimenter. z_α is the

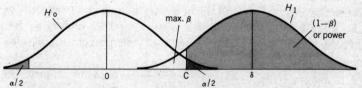

Figure 16. The curve on the left is the distribution of $\bar{X}_1 - \bar{X}_2$ when $H_o: \mu_1 - \mu_2 = 0$. The curve on the right is the distribution of $\bar{X}_1 - \bar{X}_2$ when the true $H_1: \mu_1 - \mu_2 = \delta$. The common cut-off point C is chosen so that $\alpha/2$ for the H_o curve lies above it and the maximum allowable β for the H_1 curve lies below it. *The region above C is the rejection area for both distributions.* (Since δ represents both positive and negative values of $\mu_1 - \mu_2$, the curves would be reversed for negative values.)

z score equivalent of $C - 0$ (Figure 16), when *the rejection area above C in the H_o distribution is $\alpha/2$*. z_β is the z score equivalent of $C - \delta$ when *the area of nonrejection below C in the H_1 distribution* $= \beta$ (the maximum risk of a Type II error).

This looks like Greek, but let us take two or three concrete examples and we shall see that it is relatively easy to manage.

Example A. Suppose we want to keep the risk of a false rejection of $H_o: \mu_1 - \mu_2 = 0$ fairly small, so we set $\alpha = .05$. And we decide that the power of our test to reject H_o (which may after all be false) must not be less than .80 (that is, the maximum $\beta = .20$). We also decide that in order to make our experiment meaningful, either practically or theoretically, the true minimum *absolute* value of $\mu_1 - \mu_2$ must be 3.30 (that is, $\delta = 3.30$). Working with absolute values of $\mu_1 - \mu_2$ calls for *a two-tailed test*. If $\alpha = .05$, the rejection area at one end will be $\alpha/2 = .025$. As Figure 16 shows, we are dealing with the region in which the right end of the H_o distribution overlaps the left end of the H_1 distribution. *The region above the cut-off point C is the rejection area in both distributions*. From Table 10, column 4, we find that the value of z that will leave .025 of the total area in the *right* tail of the H_o distribution (above C) is 1.96; so $z_\alpha = 1.96$. We must now find the value of z that will leave no more than .20 of the total area in the *left* tail of the H_1 distribution (below C). (Then the power to reject H_o will be at least .80.) From Table 10, column 4, we find that $z_\beta = .84$. If we now take as the population variance $\sigma^2 = 60$, as in the chemist's diet experiment with the cats, we have all the necessary ingredients to substitute in formula *10.2*: $\sigma^2 = 60$, $\delta = 3.30$, $z_\alpha = 1.96$ and $z_\beta = .84$. Therefore, we get as an *estimate* of the proper sample size

$$n = \frac{2 \times 60}{(3.30)^2}(1.96 + .84)^2 = \frac{120 \times 7.84}{10.9} = 85.4 \text{ or approx. } 86$$

This means that there should be no less than 86 *in each* of the samples if the test is to meet the specified α and β risks, and thus have the specified power (at least .80) to detect a true difference between the population means of at least 3.30 or -3.30. (It would take larger samples to detect a smaller $|\mu_1 - \mu_2|$ with the same power.) From the point of view of the developer of "fatex," this is an awful lot of cats! But it may be worthwhile to increase the size of the samples from 30 to 86 if he wants to increase his chances of proving something, *unless* he can increase his power another way, as suggested in Method III, below.

Example B. If we decide to make the risks of a Type I and a Type II error equal and set $\alpha = .10$ and the maximum allowable $\beta = .10$ (which means that the minimum power will be .90), we get from Table 10, column 4:

$$z_\alpha = 1.65 \quad \text{and} \quad z_\beta = 1.28$$

And if we keep the other conditions as they were in Example A: $\sigma^2 = 60$, and the assumed true minimum difference between means $\mu_1 - \mu_2 = 3.30$ or -3.30, then from formula *10.2* the estimate of the sample size is

$$n = \frac{2 \times 60}{(3.30)^2} (1.65 + 1.28)^2 = \frac{120 \times 8.58}{10.9} = 94.5 \text{ or } 95 \text{ approx.}$$

This would take even more cats!

Example C. In Examples A and B we assumed that the true minimum $|\mu_1 - \mu_2|$ was 3.30 for no particular reason, except that this value was the same as the obtained $\bar{X}_1 - \bar{X}_2$ in the chemist's experiment and it is not excessively large. In standard error units it is only $3.30/2.00 = 1.65$. It might be interesting to estimate the required n if we use all the conditions of Example A except that we set the minimum $|\mu_1 - \mu_2|$ at 4.00, which is just 2.0 times the standard error of the difference, $\sigma_{(\bar{X}_1 - \bar{X}_2)}$. Then with $\sigma^2 = 60$, $z_\alpha = 1.96$, and $z_\beta = .84$, we get from formula *10.2*

$$n = \frac{2 \times 60}{(4.00)^2} (1.96 + .84)^2 = \frac{120 \times 7.84}{16.0} = 58.8 \text{ or } 59 \text{ approx.}$$

This would require fewer cats.

In general, if other conditions are held constant, the larger the absolute value of the true mean difference, $|\mu_1 - \mu_2|$, the smaller the sample size required to detect it with a specified power. Before we move on to another method of increasing test power, it should be emphasized that *formula 10.2 is not a magical device for producing significant experimental results*; but if we use the sample size which it indicates, and if our other assumptions (such as equal population variances) are correct, then we should be protected against missing a true population difference (Type II error) with the specified minimum power.

Method III: Decreasing the Estimated Variance, σ^2

If we take another look at formula *10.2*, we can see at once that estimated required sample size is directly proportional to estimated population variance, σ^2. If in Example C above we had used $\sigma^2 = 30$, instead of 60, and kept everything else the same (that is, $|\mu_1 - \mu_2| = 4.00$, $z_\alpha = 1.96$ for $\alpha = .05$, and $z_\beta = .84$ for maximum $\beta = .20$), then by formula *10.2* we would have had

$$n = \frac{2 \times 30}{(4.00)^2} (1.96 + .84)^2 = 29.4 \text{ or approx. } 30$$

Under these conditions the chemist would have a significance test with his specified risk of a Type I error of only .05, but with a power of at least .80 to detect a true difference between means of not less than |4.00.| And he would need no more than the original number of cats. *But in order to cut down σ^2 he would have to improve his experimental procedure.* He would have to select his samples with greater care and match the experimental and control groups with greater care in order to reduce the variance of his samples, from which the population variance is estimated. For example, he might take extra care to make sure that his samples all come from the same breed of cats, using litter mates as much as possible (distributed equally between experimental and control groups). He might check more carefully on exercise routines, make sure that all his cats were either spayed or not spayed, and possibly (if he were psychoanalytically oriented) even check on the degree of psychosexual adjustment of his subjects.

Final Comment: In decision making in general we should always keep in mind the distinction made earlier between *statistical significance* and *importance*. If a *t* or a *z* test leads us to conclude that we have statistically significant results in an experiment, this does not necessarily mean that the results are important. Importance frequently depends upon the *size* of the probable difference between population means, $\mu_1 - \mu_2$, among other things. It is part of the process of decision making to decide what minimum difference between population means will be regarded as important. This has been implied in the procedure for estimating sample size above in section 2, Method II.

Exercises

1. (a) Give a one-sentence definition of each of the following: Type I error, Type II error, power of a test. (b) What is the relationship of power to Type II errors?

2. (a) On what four things is the power of a test dependent? (b) In what three ways can the power of a test be increased?

3. What is the essence of the procedure for calculating the power of a test? What four steps are recommended for carrying out this procedure?

4. If a decision has been made to set the risk of a Type I error at .05 in a two-tailed z test, and if the standard error of the difference between two means has been estimated to be 2.00, calculate the power of the test if the true difference between the means, $\mu_1 - \mu_2$, is 2.24. (Use the four steps called for in Exercise 3, drawing a sketch corresponding to Figure 15 for step 1.) What is the probability of a Type II error?

5. Repeat Exercise 4 assuming $\mu_1 - \mu_2$ to be 3.92.

6. If it is decided to set $\alpha = .10$ in a two-tailed z test, and if it is estimated that $\sigma_{(\bar{X}_1 - \bar{X}_2)} = 2.00$, calculate the power for an assumed $\mu_1 - \mu_2$ of 5.29. (Hint: in calculating p_r see Example 4, p. 69.) What is the risk of a Type II error?

7. (a) If two groups of 30 each have approximately equal variances and yield an estimate of the population variance of $\sigma^2 = 15$, what is the estimated standard error of the difference between the means? (b) If the risk of a Type I error is set at .20, what will be the power, using a two-tailed z test, if the true difference between means, $\mu_1 - \mu_2$, is 1.00? (c) What is the risk of a Type II error?

8. Repeat Exercise 7 assuming $\mu_1 - \mu_2 = 2.12$. (Hint: same hint as for Exercise 6.)

9. If a pilot study has suggested that a good estimate of

population variance is 32 and an experimenter has decided to set the risk of a Type I error at .05 and the maximum risk of a Type II error at .10 in a two-sided significance test, estimate the size of the sample required in each group to detect a true absolute difference, $|\mu_1 - \mu_2|$, of 2.00.

10. Repeat Exercise 9 substituting $\alpha = .10$, minimum power $= .84$, and $|\mu_1 - \mu_2| = 4.00$.

11
Simple Analysis of Variance

In the last two chapters we have dealt with tests of significance for the difference between the means of two groups. The t test (or the z test for groups large enough to base probabilities on the normal distribution) was adequate for a comparison of just two groups, for a test of a single difference between two means. If we have three independent groups, A, B, and C, we can still use the t test to compare the means of A and B, B and C, and A and C. This is, in fact, the simplest procedure. If, however, we had six groups, there would then be 15 possible comparisons between two means; and with 12 groups there would be 66 possible comparisons between two means.* It would certainly be a

---

*The formula for the number of possible comparisons of two means for k groups is

$$_kC_2 = \frac{k(k-1)}{2}$$

tedious job to make 66 t tests, even if the groups were small. And it might also be a fruitless task.

It would be very useful to have some way of knowing in advance if somewhere in a mass of possible comparisons we were likely to find one or more that might turn out to be significant. This would keep us from blindly wading into the mess, perhaps never to come up with anything of value. Well, this is precisely what simple analysis of variance does for us. *It is a kind of collective t test.* And if the preliminary analysis tells us that significant differences do exist *somewhere* among the possible comparisons, there are methods for locating more precisely just where such differences lie. Let us first, however, find out how to make this collective t test.

1. Main Steps in Simple Analysis of Variance

When we learned earlier how to calculate the variance (s^2) for a single set of scores (Chapter 4), we saw that the essence of the calculation was Σx^2, which could be determined in several ways: from deviation scores, raw scores, or coded scores. This Σx^2, which is called "the sum of squares" and is designated by the symbol SS, was later divided by $n - 1$ to get from a sample an "unbiased estimate" of the population variance, σ^2.

Now in carrying out a simple analysis of variance for k different experimental groups (often called "treatment" groups, because each receives a different treatment), there are *four main steps*.

Step 1. We first get *the total sum of squares*, SS_T, and we

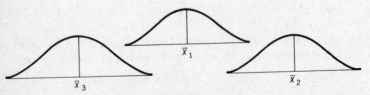

Figure 17. Two kinds of variation. The curves taken separately illustrate variation *within* groups; taken together they illustrate variation *between* group means.

analyze it, or break it down into *two major component parts*. One part is due to *variation within the groups*, SS_W, and the other is due to *variation between the means of the groups*, SS_B. Figure 17 illustrates these two kinds of variation. The procedure for calculating the three sums of squares is described in section 4 below.

Step 2. We then use SS_W and SS_B to make two independent estimates of the population variance. This is done by dividing each by an appropriate number of *degrees of freedom* (*d.f.*), indicated in the denominators of formulas *11.1* and *11.2* below. The resulting variances are called *mean squares*, and are symbolized by MS_W and MS_B. The formulas are

$$MS_W = \frac{SS_W}{k(n-1)} \qquad (11.1)$$

and

$$MS_B = \frac{SS_B}{k-1} \qquad (11.2)$$

where k is the number of experimental, or treatment, groups and n is the number of cases in each group. These calculations are illustrated in the summary of the variance analysis on page 124.

Step 3. We next calculate the F ratio. This is simply the ratio of the two estimates of the population variance:

$$F = \frac{MS_B}{MS_W} \qquad (11.3)$$

This is also illustrated in the summary on page 124.

Step 4. And, finally, we interpret our obtained value of F with the help of Table 13, which gives us the critical F values for the $p = .05$ and $p = .01$ levels of significance for over 400 selected *d.f.* combinations. Before we take up the interpretation of F for this particular kind of analysis let us get a rough idea of the general nature of the variance ratio and the table of F.

2. The General Nature of the *F* Ratio and the *F* Table

Table 13 is based not on one but on many random sampling distributions of the F ratio. The F ratio may be defined as the

TABLE 13

The $p = .05$ and $p = .01$ Points for Selected Distributions of F^a
(THE $p = .01$ VALUES ARE IN BOLDFACE TYPE.)

d.f. for larger MS	\multicolumn{11}{c}{d.f. for the smaller variance (MS)}
	8	10	12	14	16	18	20	24	28	32	36
1	5.32	4.96	4.75	4.60	4.49	4.41	4.35	4.26	4.20	4.15	4.11
	11.26	**10.04**	**9.33**	**8.86**	**8.53**	**8.28**	**8.10**	**7.82**	**7.64**	**7.50**	**7.39**
2	4.46	4.10	3.88	3.74	3.63	3.55	3.49	3.40	3.34	3.30	3.26
	8.65	**7.56**	**6.93**	**6.51**	**6.23**	**6.01**	**5.85**	**5.61**	**5.45**	**5.34**	**5.25**
3	4.07	3.71	3.49	3.34	3.24	3.16	3.10	3.01	2.95	2.90	2.86
	7.59	**6.55**	**5.95**	**5.56**	**5.29**	**5.09**	**4.94**	**4.72**	**4.57**	**4.46**	**4.38**
4	3.84	3.48	3.26	3.11	3.01	2.93	2.87	2.78	2.71	2.67	2.63
	7.01	**5.99**	**5.41**	**5.03**	**4.77**	**4.58**	**4.43**	**4.22**	**4.07**	**3.97**	**3.89**
5	3.69	3.33	3.11	2.96	2.85	2.77	2.71	2.62	2.56	2.51	2.48
	6.63	**5.64**	**5.06**	**4.69**	**4.44**	**4.25**	**4.10**	**3.90**	**3.76**	**3.66**	**3.58**
6	3.58	3.22	3.00	2.85	2.74	2.66	2.60	2.51	2.44	2.40	2.36
	6.37	**5.39**	**4.82**	**4.46**	**4.20**	**4.01**	**3.87**	**3.67**	**3.53**	**3.42**	**3.35**
7	3.50	3.14	2.92	2.77	2.66	2.58	2.52	2.43	2.36	2.32	2.28
	6.19	**5.21**	**4.65**	**4.28**	**4.03**	**3.85**	**3.71**	**3.50**	**3.36**	**3.25**	**3.18**
8	3.44	3.07	2.85	2.70	2.59	2.51	2.45	2.36	2.29	2.25	2.21
	6.03	**5.06**	**4.50**	**4.14**	**3.89**	**3.71**	**3.56**	**3.36**	**3.23**	**3.12**	**3.04**
9	3.39	3.02	2.80	2.65	2.54	2.46	2.40	2.30	2.24	2.19	2.15
	5.91	**4.95**	**4.39**	**4.03**	**3.78**	**3.60**	**3.45**	**3.25**	**3.11**	**3.01**	**2.94**

10	3.35	2.97	2.76	2.60	2.49	2.41	2.35	2.26	2.19	2.14	2.10
	5.82	4.85	4.30	3.94	3.69	3.51	3.37	3.17	3.03	2.94	2.86
11	3.31	2.94	2.72	2.56	2.45	2.37	2.31	2.22	2.15	2.10	2.06
	5.74	4.78	4.22	3.86	3.61	3.44	3.30	3.09	2.95	2.86	2.78
12	3.28	2.91	2.69	2.53	2.42	2.34	2.28	2.18	2.12	2.07	2.03
	5.67	4.71	4.16	3.80	3.55	3.37	3.23	3.03	2.90	2.80	2.72
14	3.23	2.86	2.64	2.48	2.37	2.29	2.23	2.13	2.06	2.02	1.98
	5.56	4.60	4.05	3.70	3.45	3.27	3.13	2.93	2.80	2.70	2.62
16	3.20	2.82	2.60	2.44	2.33	2.25	2.18	2.09	2.02	1.97	1.93
	5.48	4.52	3.98	3.62	3.37	3.19	3.05	2.85	2.71	2.62	2.54
20	3.15	2.77	2.54	2.39	2.28	2.19	2.12	2.02	1.96	1.91	1.87
	5.36	4.41	3.86	3.51	3.25	3.07	2.94	2.74	2.60	2.51	2.43
24	3.12	2.74	2.50	2.35	2.24	2.15	2.08	1.98	1.91	1.86	1.82
	5.28	4.33	3.78	3.43	3.18	3.00	2.86	2.66	2.52	2.42	2.35
30	3.08	2.70	2.46	2.31	2.20	2.11	2.04	1.94	1.87	1.82	1.78
	5.20	4.25	3.70	3.34	3.10	2.91	2.77	2.58	2.44	2.34	2.26
50	3.03	2.64	2.40	2.24	2.13	2.04	1.96	1.86	1.78	1.74	1.69
	5.06	4.12	3.56	3.21	2.96	2.78	2.63	2.44	2.30	2.20	2.12
100	2.98	2.59	2.35	2.19	2.07	1.98	1.90	1.80	1.72	1.67	1.62
	4.96	4.01	3.46	3.11	2.86	2.68	2.53	2.33	2.18	2.08	2.00

Table adapted by permission from STATISTICAL METHODS, 6th edition, by George W. Snedecor and William C. Cochran, © 1967 by The Iowa State University Press, Ames, Iowa.

[a]The *upper* end of the distributions only.

TABLE 13 (continued)

The $p = .05$ and $p = .01$ Points for Selected Distributions of F
(THE $p = .01$ VALUES ARE IN BOLDFACE TYPE.)

d.f. for the smaller variance (MS)

d.f. for larger MS	42	48	55	65	80	100	125	150	200	400	1000
1	4.07	4.04	4.02	3.99	3.96	3.94	3.92	3.91	3.89	3.86	3.85
	7.27	**7.19**	**7.12**	**7.04**	**6.96**	**6.90**	**6.84**	**6.81**	**6.76**	**6.70**	**6.66**
2	3.22	3.19	3.17	3.14	3.11	3.09	3.07	3.06	3.04	3.02	3.00
	5.15	**5.08**	**5.01**	**4.95**	**4.88**	**4.82**	**4.78**	**4.75**	**4.71**	**4.66**	**4.62**
3	2.83	2.80	2.78	2.75	2.72	2.70	2.68	2.67	2.65	2.62	2.61
	4.29	**4.22**	**4.16**	**4.10**	**4.04**	**3.98**	**3.94**	**3.91**	**3.88**	**3.83**	**3.80**
4	2.59	2.56	2.54	2.51	2.48	2.46	2.44	2.43	2.41	2.39	2.38
	3.80	**3.74**	**3.68**	**3.62**	**3.56**	**3.51**	**3.47**	**3.44**	**3.41**	**3.36**	**3.34**
5	2.44	2.41	2.38	2.36	2.33	2.30	2.29	2.27	2.26	2.23	2.22
	3.49	**3.42**	**3.37**	**3.31**	**3.25**	**3.20**	**3.17**	**3.14**	**3.11**	**3.06**	**3.04**
6	2.32	2.30	2.27	2.24	2.21	2.19	2.17	2.16	2.14	2.12	2.10
	3.26	**3.20**	**3.15**	**3.09**	**3.04**	**2.99**	**2.95**	**2.92**	**2.90**	**2.85**	**2.82**
7	2.24	2.21	2.18	2.15	2.12	2.10	2.08	2.07	2.05	2.03	2.02
	3.10	**3.04**	**2.98**	**2.93**	**2.87**	**2.82**	**2.79**	**2.76**	**2.73**	**2.69**	**2.66**
8	2.17	2.14	2.11	2.08	2.05	2.03	2.01	2.00	1.98	1.96	1.95
	2.96	**2.90**	**2.85**	**2.79**	**2.74**	**2.69**	**2.65**	**2.62**	**2.60**	**2.55**	**2.53**
9	2.11	2.08	2.05	2.02	1.99	1.97	1.95	1.94	1.92	1.90	1.89
	2.86	**2.80**	**2.75**	**2.70**	**2.64**	**2.59**	**2.56**	**2.53**	**2.50**	**2.46**	**2.43**

10	2.06 2.77	2.03 2.71	2.00 2.66	1.98 2.61	1.95 2.55	1.92 2.51	1.90 2.47	1.89 2.44	1.87 2.41	1.85 2.37	1.84 2.34
11	2.02 2.70	1.99 2.64	1.97 2.59	1.94 2.54	1.91 2.48	1.88 2.43	1.86 2.40	1.85 2.37	1.83 2.34	1.81 2.29	1.80 2.26
12	1.99 2.64	1.96 2.58	1.93 2.53	1.90 2.47	1.88 2.41	1.85 2.36	1.83 2.33	1.82 2.30	1.80 2.28	1.78 2.23	1.76 2.20
14	1.94 2.54	1.90 2.48	1.88 2.43	1.85 2.37	1.82 2.32	1.79 2.26	1.77 2.23	1.76 2.20	1.74 2.17	1.72 2.12	1.70 2.09
16	1.89 2.46	1.86 2.40	1.83 2.35	1.80 2.30	1.77 2.24	1.75 2.19	1.72 2.15	1.71 2.12	1.69 2.09	1.67 2.04	1.65 2.01
20	1.82 2.35	1.79 2.28	1.76 2.23	1.73 2.18	1.70 2.11	1.68 2.06	1.65 2.03	1.64 2.00	1.62 1.97	1.60 1.92	1.58 1.89
24	1.78 2.26	1.74 2.20	1.72 2.15	1.68 2.09	1.65 2.03	1.63 1.98	1.60 1.94	1.59 1.91	1.57 1.88	1.54 1.84	1.53 1.81
30	1.73 2.17	1.70 2.11	1.67 2.06	1.63 2.00	1.60 1.94	1.57 1.89	1.55 1.85	1.54 1.83	1.52 1.79	1.49 1.74	1.47 1.71
50	1.64 2.02	1.61 1.96	1.58 1.90	1.54 1.84	1.51 1.78	1.48 1.73	1.45 1.68	1.44 1.66	1.42 1.62	1.38 1.57	1.36 1.54
100	1.57 1.91	1.53 1.84	1.50 1.78	1.46 1.71	1.42 1.65	1.39 1.59	1.36 1.54	1.34 1.51	1.32 1.48	1.28 1.42	1.26 1.38

ratio of two independent, unbiased estimates of population variance. If we repeatedly make such estimates from two samples taken at random (by chance), one from each of two normally distributed populations with the same variance, the resulting F ratios will constitute a random variable. The shape of the frequency distribution curves for F varies considerably with the size of the samples (and the related degrees of freedom) used in making the unbiased estimates. Though F may range from zero to infinity, most F distributions have their maximum frequencies in the neighborhood of 1.00 (and are consequently rather skewed). Values far above 1.00, or close to zero, are relatively less frequent, or *less probable*, than those nearer 1.00.

Since the samples used in estimating the two variances in the F ratio do not have to be of the same size, the number of possible combinations of sample size, and the related $d.f.$'s, is infinite. This means that, in theory at least, there is an infinite number of somewhat different F distributions. Consequently, a table of F, like Table 13, must be selective in at least two ways. It involves F distributions for only selected combinations of degrees of freedom that are commonly useful; and for each of these distributions (over 400) it lists only the critical $p = .05$ and $p = .01$ F values. Furthermore, since in analysis of variance we are interested only in values of F which are larger than 1.00, the lower parts of the distributions, between 1.00 and zero, are omitted altogether. So what we have in Table 13 is information for *a one-tailed significance test only*. But, as we shall see shortly, this is what we want.

3. Interpretation of the F Ratio

The null hypothesis we are testing is that the k groups are merely random samples from the same normally distributed population of the variable X, which has a mean μ and a variance σ^2 (or from populations with the same μ's and σ^2's). If this hypothesis were true, the *average* values of the k sample means would, in the long run, all be equal to μ. In symbols, $H_o: \mu_1 = \mu_2 = \ldots \mu_k = \mu$. This implies that there are no experimental effects. It does *not* imply, however, that the *individual* sample means have to be equal, for even if the different experimental conditions produce no measurable differences in average performance, the sample means will exhibit the usual *chance* variations about the population mean. *We retain the null hypothesis as probably correct if the F ratio is not significantly greater than 1.00*, which means that the estimate of the population variance based on the variation *between* the sample means, MS_B, is not essentially different from the estimate based on the variation *within* the samples, MS_W. *And we conclude that there are no experimental effects; that is, there are no significant differences between any two means in the k groups.*

If, however, the F ratio is found to be significantly greater than 1.00 (as indicated for the appropriate d.f.'s in Table 13), MS_B is significantly greater than MS_W, and the null hypothesis is probably false: the k samples probably did not all come from the same population. The *extra* variability in the MS_B estimate is probably due to *real* differences between the groups rather than to *chance* variations in sampling. In other words, *we conclude that there is probably a significant difference for at least one comparison of means somewhere in the k groups. There may be several significant differences.*

Before we deal with methods for spotting the location of such possible experimental effects, let us see how we determine the three sums of squares (SS's) which are involved in the calculation of the F ratio (Step 1, section 1, above).

4. Calculation of the Three Sums of Squares for Simple Analysis of Variance

We shall use to illustrate this procedure the *simplified data* for just three groups given in Table 14. This data might come from an experiment designed to test the relative reinforcing effect of some form of "reward" and some form of "punishment," verbal or otherwise, as indicated by the variable X.

The formulas involved in these calculations look more formidable than they are, because of the subscripts and superscripts required to avoid ambiguity. It is helpful to observe that

TABLE 14

Simplified Data for $k = 3$ Treatment Groups, with $n = 5$ Subjects in Each Group Selected at Random

	Punishment (1)		Reward (2)		Control (3)	
	X_1	X_1^2	X_2	X_2^2	X_3	X_3^2
	3	9	5	25	1	1
	5	25	7	49	3	9
	4	16	6	36	2	4
	6	36	8	64	4	16
	2	4	4	16	0	0
Σ's	20	90	30	190	10	30
	$\bar{X}_1 = \dfrac{20}{5} = 4.0$		$\bar{X}_2 = \dfrac{30}{5} = 6.0$		$\bar{X}_3 = \dfrac{10}{5} = 2.0$	

they are all based on the familiar formula for the sum of squares (SS) which makes use of original scores:

$$SS = \Sigma x^2 = [\Sigma(X - \bar{X})^2] = \Sigma X^2 - \frac{(\Sigma X)^2}{n} \quad (4.4)$$

They all have the same *general* pattern.

A. Total SS

$$SS_T = \left[\sum_1^{kn}(X - \bar{X}_{kn})^2\right] = \sum_1^{kn} X^2 - \frac{\left(\sum_1^{kn} X\right)^2}{kn} \quad (11.4)$$

where $\bar{X}_{kn}$ is the "grand mean" of all n cases in all k groups, and Σ_1^{kn} stands for "the sum of" the indicated variable from the first case through all n cases in all k groups. In other words, kn covers all $k \times n$ cases. *The quantity in the brackets does not have to be calculated*; it is given to help explain the rest of the formula. Hence, for the data of Table 14

$$SS_T = (90 + 190 + 30) - \frac{(20 + 30 + 10)^2}{3 \times 5}$$
$$= 310 - 240 = 70$$

B. SS within Groups

$$SS_W = \sum_1^k \sum_1^n (X_k - \bar{X}_k)^2$$

where the *subscript* k takes values 1, 2, ... k \quad (11.5)

The first summation sign (Σ_1^k) simply indicates that we must sum for all k groups the sums of squares called for by the second summation sign. Thus, working out for each group the SS's called for by the second Σ, we get

For Group 1:

$$SS_{W1} = \left[\sum_1^n (X_1 - \bar{X}_1)^2 \right] = \Sigma X_1^2 - \frac{(\Sigma X_1)^2}{n}$$

$$= 90 - \frac{(20)^2}{5} = 90 - 80 = 10$$

For Group 2:

$$SS_{W2} = \left[\sum_1^n (X_2 - \bar{X}_2)^2 \right] = \Sigma X_2^2 - \frac{(\Sigma X_2)^2}{n}$$

$$= 190 - \frac{(30)^2}{5} = 190 - 180 = 10$$

For Group 3:

$$SS_{W3} = \left[\sum_1^n (X_3 - \bar{X}_3)^2 \right] = \Sigma X_3^2 - \frac{(\Sigma X_3)^2}{n}$$

$$= 30 - \frac{(10)^2}{5} = 30 - 20 = 10$$

(Again, the quantities in brackets do not have to be calculated; they are merely explanatory.)

Then, getting the sum of these group SS's, we have

$$SS_W = 10 + 10 + 10 = 30$$

C. SS between Groups

$$SS_B = n \left[\sum_1^k (\bar{X}_k - \bar{X}_{kn})^2 \right]$$

$$= n \left[\sum_1^k \bar{X}_k^2 - \frac{\left(\sum_1^k \bar{X}_k \right)^2}{k} \right] \qquad (11.6)$$

where $\bar{X}_{kn}$ is the "grand mean" of all the cases (as it was in A above) or the mean of the k means (which amounts to the same thing); and $\bar{X}_k$ represents the individual means of the k groups as k takes the successive values $1, 2, \ldots k$. When we realize that *the basic unit in this formula is a group mean* ($\bar{X}_k$), rather than an individual score (X), we can see that the pattern of the part in brackets is the same as that of the familiar formula *4.4*. The quantities in brackets are multiplied by n because each $\bar{X}_k$ is based on n cases. *The first quantity in brackets does not have to be calculated*; it is given to help explain the rest of the formula. Hence, for the data of Table 14.

$$SS_B = 5\left[(4.0^2 + 6.0^2 + 2.0^2) - \frac{(4.0 + 6.0 + 2.0)^2}{3}\right]$$
$$= 5\left[56 - \frac{144}{3}\right] = 5 \times 8 = 40$$

(*Note:* an alternative procedure for calculating SS_B is given on page 135.)

We can now check our three SS calculations. SS_T should equal $SS_W + SS_B$. It does: $70 = 30 + 40$. This takes care of *Step 1*, section 1. The other three steps are covered in the next section.

5. Summary of Simple Analysis of Variance; Calculation and Significance of F

Steps 2 and 3 of section 1 are so simple that they can be included in the summary of the analysis below.

Summary of Simple Analysis of Variance for the Data of Table 14

Source of Variance	SS	d.f.	MS	$F (= MS_B/MS_W)$
Between groups	40	$k - 1 = 2$	$40/2 = 20$	$20/2.5 = 8.0$
Within groups	30	$k(n - 1) = 12$	$30/12 = 2.5$	
Total	70	$kn - 1 = 14$	(check)	

Step 4 deals with the interpretation of the obtained F ratio. Does it indicate a probable significant difference between two means anywhere in our k groups? The principles for interpreting an obtained value of F were elaborated in section 3. The null hypothesis is that the k experimental groups are all random samples from the same population, or from populations with the same mean, μ, and the same variance, σ^2. The distributions of F are based on similar assumptions. *The table of F (p. 116)* gives us the $p = .05$ and $p = .01$ values of the F ratio for selected combinations of degrees of freedom. In this case the row labels are the *d.f.* values for MS_B and the column headings are the *d.f.* values for MS_W.

In our illustrative experiment we have for these two mean squares 2 and 12 degrees of freedom, respectively. From the table we find in the row for 2 *d.f.* and the column for 12 *d.f.* that $F = 3.88$ for $p = .05$ and 6.93 for $p = .01$. Our obtained value of F was 8.00, which is even larger than 6.93. This means that, on the basis of our null hypothesis, the probability of getting a value of F this large is less than .01. *So we reject H_o with a high degree of confidence and conclude that there is probably at least one significant difference among the possible comparisons of two means in our k groups.* This may be put more succinctly: Since $F = 8.00 > 6.93$, we may reject H_o at the $p < .01$ level and conclude that there is a highly significant overall experimental effect.

The next problem is to find out just where this experimental effect is located. A relatively simple procedure for doing this is described in the next section.

6. The Scheffé Test of Significance for a Difference between Any Two Means

The Scheffé significance test is in effect a generalized t test. We recall that the main effort in applying the t test to a difference between two means goes into calculating *the standard error of the difference*, $s_{(\bar{X}_1-\bar{X}_2)}$. This appears in the denominator of the t ratio:

$$t = \frac{\text{Diff.}}{s_{\text{Diff.}}} = \frac{\bar{X}_1 - \bar{X}_2}{s_{(\bar{X}_1-\bar{X}_2)}} \qquad (8.1)$$

A common formula for the standard error of the difference, which applies when n is the same for the two samples, is

$$s_{(\bar{X}_1-\bar{X}_2)} = \sqrt{\frac{2s^2}{n}} \qquad (8.3)$$

where

$$s^2 = \frac{\Sigma x_1^2 + \Sigma x_2^2}{n_1 + n_2 - 2}$$

We see from these formulas that quite a bit of arithmetical labor may go into calculating a single t ratio. If we had to calculate t's for a large number of possible comparisons of two means, this would be exhausting and possibly irritating.

Now the Scheffé method saves a great deal of time and effort by making use of *the same standard error of the difference*, $s_{Diff.}$, *for all comparisons of two means in the k groups*. This common $s_{Diff.}$ is based on an estimate of the population variance which takes into account the variability in all k samples, not just two. That is, for s^2 in formula *8.3* we substitute MS_W, which we have already calculated. The formula becomes

$$s_{Diff.} = \sqrt{\frac{2MS_W}{n}} \qquad (11.7)$$

This may seem like cheating; but it isn't. Actually, an estimate based on all k samples should be *more reliable* (if our null hypothesis assumption of a common population is correct) than one based on only two random samples, which is the case in formula *8.3*. Using the results from our analysis of variance summary (p. 124), we get, without pain, for the common standard error of the difference

$$s_{Diff.} = \sqrt{\frac{2MS_W}{n}} = \sqrt{\frac{2 \times 2.5}{5}} = 1.0$$

For *the criterion of significance* the Scheffé method uses t'. This is obtained from the formula

$$t' = \sqrt{(k-1)F'} \qquad (11.8)$$

where F' is the tabled value of F for $k - 1$ degrees of freedom for the larger MS and $k(n - 1)$ d.f. for the smaller MS. If the *absolute* value of an obtained t is equal to, or greater than t', the difference between the means is considered significant. In our three-group experiment, with $n = 5$ in each group, we have already found that, for 2 and 12 d.f., $F' = 3.88$ for $p = .05$ and 6.93 for $p = .01$. Substituting these values, one at a time, in the formula above, we get

$$t'_{.05} = \sqrt{2 \times 3.88} = 2.79$$

and

$$t'_{.01} = \sqrt{2 \times 6.93} = 3.72$$

It is now an extremely simple matter to make the only three significance tests between two means that are possible when $k = 3$. In our experiment

$$t_{12} = \frac{\bar{X}_1 - \bar{X}_2}{s_{\text{Diff.}}} = \frac{4.0 - 6.0}{1.0} = -2.0,$$

which has an *absolute* value of 2.0.

Since $2.0 < 2.79$, we do not reject $H_o: \mu_1 = \mu_2$ and we conclude that the difference between the "punishment" and "reward" groups is not significant at even the $p = .05$ level.

$$t_{23} = \frac{\bar{X}_2 - \bar{X}_3}{s_{\text{Diff.}}} = \frac{6.0 - 2.0}{1.0} = 4.0$$

Since $4.0 > 3.72$, we conclude that $\mu_2 \neq \mu_3$ and that the difference between the "reward" and the "control" groups is significant at the $p < .01$ level.

$$t_{13} = \frac{\bar{X}_1 - \bar{X}_3}{s_{\text{Diff.}}} = \frac{4.0 - 2.0}{1.0} = 2.0$$

Since $2.0 < 2.79$, we do not reject $H_o: \mu_1 = \mu_3$ and we conclude that the difference between the "punishment" and "control" groups is not significant even at the $p = .05$ level.

Clearly this procedure is a big time-saver even when only three tests of significance have to be made. If we had to consider as many as 45 possible comparisons of two means each (which would be the case for $k = 10$ groups), we would be way ahead using the Scheffé method.

As a matter of fact, *we can do even better. By taking one more step we can save 45!* Taking advantage of Scheffé's uniform standard error of the difference, $s_{\text{Diff.}}$, and t' criterion, *we can easily determine the minimum difference between any two means required for significance.* This will be

$$\text{Min. Diff. for Sig.} = t' s_{\text{Diff.}} \qquad (11.9)*$$

*Scheffé is not responsible for this formula, but it seems a logical corollary of his method.

For example, if we have $k = 10$ groups, with $n = 10$ in each, and the MS_W has been found to be 20, then

$$s_{\text{Diff.}} = \sqrt{\frac{2MS_W}{n}} = \sqrt{\frac{2 \times 20}{10}} = 2.0$$

And $t' = \sqrt{(k-1)F'}$, where F' is the tabled value of F for $k - 1 = 9$ and $k(n - 1) = 90$ degrees of freedom. We enter the table from the side on the row for 9 $d.f.$ and we easily interpolate *by inspection* for 90 $d.f.$ between the 80 and 100 columns. We get $F' = 1.98$ for $p = .05$ and 2.61 for $p = .01$. Then $t'_{.05} = \sqrt{9 \times 1.98} = 4.22$ and $t'_{.01} = 3\sqrt{2.61} = 8.08$. Substituting in formula *11.9*, we get

Min. Diff. for Sig.$_{.05}$ = 4.22 × 2.0 = 8.44

and

Min. Diff. for Sig.$_{.01}$ = 8.08 × 2.0 = 16.16

With this information we can tell at once whether the difference between any two specific means is significant, at either the $p = .05$ or the $p = .01$ level.

So far we have used the Scheffé procedure, and its corollary above, for comparisons involving just two means. It was designed to deal also with more complicated comparisons, such as *the average* of two or more means compared with one or more other means. The application of the method to such comparisons is described in the next section.

7. The Scheffé Test for More Complex Comparisons*

In the simplified experiment on incentives for learning that we have used throughout this chapter we had just three treatment groups. This made possible only three comparisons between two means. Only one of these turned out to be significant: the difference between the means for the "reward" and "control" groups. It is conceivable that the experimenter might be interested in comparing the average of the two incentive group means with the control group mean. By a simple extension and

---

*This section may be omitted at the discretion of the instructor.

slight modification of what has gone before, the Scheffé method can be applied to this and other more complicated comparisons. There are *two principal features of the extended procedure*. One involves *a restriction on the kinds of comparison that can properly be made*. The other involves *a modification of the formula for the standard error of the difference* which was used when only two means were compared. Both involve the coefficients of the means being compared. This is how it works.

A. *Legitimate Comparisons Involving k Means*

The three comparisons between means that were made in the significance tests in the section above could have been written:

$$\bar{X}_1 - \bar{X}_2 = (1)\bar{X}_1 + (-1)\bar{X}_2 + (0)\bar{X}_3 \qquad (C_1)$$

$$\bar{X}_2 - \bar{X}_3 = (0)\bar{X}_1 + (1)\bar{X}_2 + (-1)\bar{X}_3 \qquad (C_2)$$

$$\bar{X}_1 - \bar{X}_3 = (1)\bar{X}_1 + (0)\bar{X}_2 + (-1)\bar{X}_3 \qquad (C_3)$$

Yes, but why? Why take a perfectly simple expression for the difference between two means and make it look so complicated? The reason is that when the k means have meaningful coefficients in front of them, we can easily apply *the legitimacy test* to the comparison. With the proper coefficients written in, the test is very simple: *The sum of the coefficients must equal zero.* In symbols, if we let a_k represent the various coefficients,

$$\Sigma a_k = 0, \text{ where } k \text{ takes values } 1, 2, \ldots k \qquad (11.10)$$

The three comparisons above are all clearly respectable. In the case of such simple comparisons, involving the difference between only two means, it is not necessary to write in the coefficients, because it is obvious that $1 - 1 = 0$. But we should write them in for more complicated comparisons. For example, the comparison of *the average* of the first two means with the third mean would be written

$$(\tfrac{1}{2})\bar{X}_1 + (\tfrac{1}{2})\bar{X}_2 + (-1)\bar{X}_3 \qquad (C_4)$$

This is a proper comparison because $\tfrac{1}{2} + \tfrac{1}{2} - 1 = 0$. And, if we had six groups and wanted to compare the average of the first two means with the average of the next three and leave the sixth out of it, we would write

$$(\tfrac{1}{2})\bar{X}_1 + (\tfrac{1}{2})\bar{X}_2 + (-\tfrac{1}{3})\bar{X}_3 + (-\tfrac{1}{3})\bar{X}_4 \qquad (C_5)$$
$$+ (-\tfrac{1}{3})\bar{X}_5 + (0)\bar{X}_6$$

This is also a proper comparison because $\tfrac{1}{2} + \tfrac{1}{2} - \tfrac{1}{3} - \tfrac{1}{3} - \tfrac{1}{3} + 0 = 0$.

B. *The Standard Error of a Comparison*

In Scheffé's generalized significance test for a difference between *any two means* in k groups of n each the standard error of the difference was given by the formula

$$s_{\text{Diff.}} = \sqrt{\frac{2MS_W}{n}} \qquad (11.7)$$

When we work with more complex comparisons, the formula for *the standard error of a comparison* is a modification of this formula

$$s_C = \sqrt{\frac{MS_W}{n} \cdot \Sigma a_k^2} \qquad (11.11)$$

The 2 in formula *11.7* has been replaced by

$$\Sigma a_k^2 \qquad (11.12)$$

where a_k represents the various coefficients in the comparison and k takes the values $1, 2, \ldots k$. Using the coefficients from comparison C_4, we get

$$\Sigma a_k^2 = (\tfrac{1}{2})^2 + (\tfrac{1}{2})^2 + (-1)^2 = \frac{3}{2}$$

This constant can be used in calculating the standard error for any comparison of three means that has the same general pattern; that is, one which involves the average of two means compared with the third. (The constant which goes with comparison C_5 would be 5/6.)

Now going back to our original illustrative experiment, let us see if the average of the "punishment" and "reward" group means is significantly different from the "control" group mean. The following data and results are already available: $\bar{X}_1 = 4.0$,

$\bar{X}_2 = 6.0$, $\bar{X}_3 = 2.0$, $MS_W = 2.5$, $n = 5$, and $\Sigma a_k^2 = \frac{3}{2}$ for comparison C_4. The standard error of this comparison we get from formula *11.11*:

$$s_{C_4} = \sqrt{\frac{2.5}{5} \times \frac{3}{2}} = \frac{\sqrt{3}}{2} = .866$$

The t ratio for a comparison is

$$t_C = \frac{C}{s_C} \qquad (11.13)$$

In this case for comparison C_4

$$t_{C_4} = \frac{C_4}{s_{C_4}} = \frac{\frac{\bar{X}_1}{2} + \frac{\bar{X}_2}{2} - \bar{X}_3}{s_{C_4}} = \frac{2.0 + 3.0 - 2.0}{.866} = \frac{3.0}{.866} = 3.46$$

The criterion of significance remains the same

$$t' = \sqrt{(k-1)F'} = 2.79 \text{ for } p = .05$$

and

$$t' = 3.72 \text{ for } p = .01 \text{ (page 126)}$$

Since $3.72 > 3.46 > 2.79$, we may conclude that the average of the two incentive groups is significantly better than the control group at the $p < .05$ level but not at the $p < .01$ level.

Exercises

1. Thirty subjects are divided at random into three groups of 10 each. Each group is subjected to a different experimental treatment. The performance of the groups on test variable X is summarized below:

Group	1	2	3
ΣX	20	40	60
ΣX^2	60	180	440
$\bar{X}$	2.0	4.0	6.0

(a) Find SS_T. (b) Find SS_W. (c) Find SS_B. (d) How many *d.f.* should be used in calculating MS_W? How many for MS_B? (e) Find MS_W, MS_B, and F. (f) Interpret F, giving H_o and conclusion.

2. Twenty subjects are assigned at random to four treatment groups of five each. Their performance on test variable X is given below:

$$\begin{array}{cccc} X_1 & X_2 & X_3 & X_4 \\ 0 & 2 & 10 & 3 \\ 2 & 4 & 9 & 6 \\ 4 & 5 & 6 & 5 \\ 1 & 3 & 8 & 7 \\ 3 & 6 & 7 & 9 \end{array}$$

Is there a significant overall experimental effect? [Hint: after the necessary preliminary steps, follow steps (a) to (f) of Exercise 1.]

3. Apply the Scheffé significance test to all possible differences between two means in Exercise 1 using the following steps: (a) Find the common s_{Diff}. (b) Find $t'_{.05}$ and $t'_{.01}$. (c) Find *Min. Diff. for Sig.* for both $p = .05$ and $.01$. (d) What differences, if any, are significant? At what level?

4. Record the four means for the data of Exercise 2. Then apply the Scheffé test to all possible differences between two means, using steps (a) to (d) of Exercise 3.

5. Apply the Scheffé test for more complex comparisons to the data of Exercise 1 to see if $\bar{X}_3$ is significantly greater than the average of $\bar{X}_1$ and $\bar{X}_2$, using the following steps: (a) Set up the comparison with the proper coefficients and show that it is "legitimate." (b) Find s_{Diff}. (c) Find t. (d) Find $t'_{.05}$ and $t'_{.01}$. (e) What is your conclusion? At what level?

6. Using the means from Exercise 2 and the Scheffé test, see if the average of $\bar{X}_3$ and $\bar{X}_4$ is significantly greater than the average of $\bar{X}_1$ and $\bar{X}_2$. Use steps (a) to (e) of Exercise 5.

7. Using the means from Exercise 2 and the Scheffé test, see if $\bar{X}_3$ is significantly greater than the average of $\bar{X}_1$, $\bar{X}_2$, and $\bar{X}_4$. Use steps (a) to (e) of Exercise 5.

12
Two-Factor Analysis of Variance

In the preceding chapter we used analysis of variance as a kind of collective t test. The technique could be applied to three or more independent experimental, or treatment, groups. Only one experimental variable, or factor, was involved in each group. In this chapter we shall use experimental groups which involve in each case two variables, or factors. We shall be interested, not only in the effect of each of the experimental variables in themselves, but in the possible *interaction* between them.

In the simple analysis of variance of the last chapter there was *one main stage*, which we broke down into four steps. The four steps involved the calculation and evaluation of the F ratio. If the F ratio indicated a significant experimental effect somewhere between the group means, the main stage was followed by the Scheffé t-test methods in order to spot the location of significant comparisons between groups.

In the two-factor analysis there are *two main stages* before similar follow-up procedures are used. *The first stage is essentially the same as the main stage in the simple analysis.* It breaks the total sum of squares into the sum of squares between groups and the sum of squares within groups; and it requires the calculation of an F ratio. It is only after the variance between groups has been shown to be significantly greater than the variance within groups that the second stage is made use of. *In the second stage the variation between group means is itself broken down into three component parts*: the variance due to each of the two experimental factors and the variance due to *interaction* between them.

In a two-factor experimental design each of the two variables, or factors, is represented in two or more ways, degrees, or "levels." There may be many levels. To illustrate the procedure, however, we shall use a simple experimental design with only two levels of each factor, "a 2×2 factorial experiment." Let us suppose that a small school system with a limited budget is considering a new textbook to be used in fifth-grade arithmetic. Before investing in the proposed text an experiment on the relative efficacy of the new and old books is run with small samples of both boys and girls. There are four test groups of five pupils each, all taught by the same teacher, two boys' groups and two girls' groups. The possible new book is used with one group of each sex and the old book is used with the other. The groups are matched for age, intelligence, and initial arithmetical ability. They are tested in arithmetic both at the beginning and the end of the experimental period. In Table 15 the variable X indicates the *improvement* of each of the subjects during the trial period. The school authorities are interested in three things: Is the proposed new text significantly more effective in general than the old one? Are fifth-grade boys generally better than fifth-grade girls in arithmetic, as taught by an experienced teacher using either textbook? Or is there an interaction effect? *An interaction effect* would imply that one text might be better for boys and the other for girls. Obviously the samples in this experiment are too small to give definitive answers to any of these questions; but we shall use the small samples to simplify the illustration of the variance analysis. This involves *two main stages*.

1. Stage I. Preliminary Analysis

This is very similar to the four main steps used in the last chapter; for the treatment combinations may be thought of as four independent samples selected at random from a common population. We can, however, make *a few changes in the calculation procedure* in order to save time, now that we understand what we were doing in simple analysis of variance.

The first change is simply the substitution of the symbols ab for k in the formulas; for if each of the a levels of Factor A is matched with each of the b levels of Factor B there will be $k = a \times b$ possible combinations or experimental groups. (See Table 15.) For example, formula *11.4* for the total sum of squares becomes

$$SS_T = \sum_1^{abn} X^2 - \frac{\left(\sum_1^{abn} X\right)^2}{abn} \qquad (12.1)$$

The second change is the use of *an alternative to formula 11.6 for the sum of squares between means*, SS_B (p. 123). If we work with the second bracket of this formula and multiply both terms by n, as indicated, and then substitute ab for k, as suggested above; and if, finally, we make use of the identity

$$\bar{X}_{ab} = \frac{\sum_1^n X_{ab}}{n}$$

we get

$$SS_B = \sum_1^{ab} \frac{\left(\sum_1^n X_{ab}\right)^2}{n} - \frac{\left(\sum_1^{abn} X\right)^2}{abn} \qquad (12.2)$$

where $\Sigma_1^n X_{ab}$ is the sum of the n scores in any one of the ab groups, and Σ_1^{ab} indicates that the expression to which it applies must be summed over all possible combinations of a and b

TABLE 15

Simplified Scores, Means, and Sums for a 2×2 Factorial Experiment Factor A is the Sex of Subjects, Factor B is Textbooks

	Factor Combinations ($n = 5$ in each)			
	Boys (A_1)		Girls (A_2)	
	A_1B_1 (new)	A_1B_2 (old)	A_2B_1 (new)	A_2B_2 (old)
	X_{11}	X_{12}	X_{21}	X_{22}
	3	5	7	1
	5	7	8	2
	4	6	9	4
	6	8	6	3
	2	4	10	0
$\sum_{1}^{n} X_{ab}$	20 +	30 +	40 +	10 = $\sum_{1}^{abn} X = 100$
$\bar{X}_{ab}$	4.0	6.0	8.0	2.0
$\sum_{1}^{n} X^2_{ab}$	90 +	190 +	330 +	30 = $\sum_{1}^{abn} X^2 = 640$

(that is, for all *ab* groups). This looks more hair-raising than the formula we started with, but it has one important soothing feature. *The last term is identical with the last term in the SS_T formula.* Since we start the analysis by calculating SS_T, this will save us some time. This last term, sometimes called *the correction term*, will also appear in the second stage of our two-factor analysis (section 2) and this will save us still more calculation time.

The third change is a little risky. *It is not necessary* and should not be made unless we have worked long enough with the other calculation procedures to be reasonably confident of our accuracy. *This change eliminates the direct calculation of SS_W* and obtains it indirectly by subtraction:

$$SS_W = SS_T - SS_B \qquad (12.3)$$

Step 1. Now let us go through *the calculation of the three sums of squares* with the entries in Table 15, *making use of the three changes just outlined.* This corresponds to *Step 1* of Chapter 11.

Stage I. Preliminary Analysis

The total sum of squares by formula *12.1* is

$$SS_T = \sum_1^{abn} X^2 - \frac{\left(\sum_1^{abn} X\right)^2}{abn}$$

$$= (90 + 190 + 330 + 30) - \frac{(100)^2}{2 \times 2 \times 5}$$

$$640 - 500 = 140$$

The sum of squares between means. The *first term* of formula *12.2* is

$$\sum_1^{ab} \frac{\left(\sum_1^n X_{ab}\right)^2}{n} = \frac{(20)^2 + (30)^2 + (40)^2 + (10)^2}{5} = 600$$

The *last term* of formula *12.2* is the same as the last term for SS_T:

$$\frac{\left(\sum_1^{abn} X\right)^2}{abn} = 500$$

So

$$SS_B = \sum_1^{ab} \frac{\left(\sum_1^n X_{ab}\right)^2}{n} - \frac{\left(\sum_1^{abn} X\right)^2}{abn} = 600 - 500 = 100$$

The sum of squares within the four treatment combinations is by formula *12.3*

$$SS_W = SS_T - SS_B = 140 - 100 = 40$$

We have now completed Step 1 of the preliminary analysis. *The other three steps are just the same as those for simple analysis of variance. Steps 2 and 3* are covered in the following summary.

Summary of Preliminary Analysis for the Four Treatment Combinations of Table 15, a 2×2 Factorial Experiment with $n = 5$

Source of Variance	SS	d.f.	MS	F
Between combinations	100	$ab - 1 = 3$	$100/3 = 33.3$	MS_B/MS_W
Within combinations	40	$ab(n-1) = 16$	$40/16 = 2.5$	$= 33.3/2.5 = 13.3$
Total	140	$abn - 1 = 19$	(check)	

Step 4 involves *the interpretation* of the obtained F ratio. Entering Table 13 in the row for $d.f. = 3$, we find in the column for $d.f. = 16$ that $F = 5.29$ for $p = .01$. Since $13.3 > 5.29$ (much greater), we conclude that highly significant overall experimental effects are indicated. In the second stage of the analysis we shall try to find out how much of these effects is due to Factor A (the boy–girl factor), how much to Factor B (the textbook factor), and how much to some interaction between them.

2. Stage II. Further Analysis of the Variance between Means

In the preliminary analysis we had convincing evidence that the variance between the means of the experimental groups, or factor combinations, was significantly greater than the variance within the groups. We shall now break down this variation between the group means into three component parts: that which is due to each of the two factors, A and B, and that which is due to a possible interaction between them. Interaction implies that they reinforce or neutralize each other in some way at some points.

We shall, as in the preliminary analysis, work first with

TABLE 16

Sums for the Four Factor Combinations of Table 15 Arranged in a 2 × 2 Table ($n = 5$ for each combination)

Sex	Textbooks		Row Sums
	B_1 (new)	B_2 (old)	$\sum X_a$ (for $bn = 10$)
A_1 (boys)	20	30	50
A_2 (girls)	40	10	50
Column Sums $\sum X_b$ (for $an = 10$)	60	40	$100 = \sum_{1}^{abn} X$

Note: Each of the four *original* sums (from which the row and column sums are obtained) was symbolized by $\sum_{1}^{n} X_{ab}$ in Table 15.

various sums of squares (*SS*'s) and then calculate the variances (*MS*'s) from them by dividing by the appropriate number of degrees of freedom. This can be done most conveniently by rearranging the group sums ($\Sigma_1^n X_{ab}$'s) from Table 15 in the form of a 2 × 2 table. This has been done in Table 16 with the two levels of Factor *A* (boys and girls) in the rows and the two levels of Factor *B* (new and old textbooks) in the columns. *There is an extra column on the right which gives the sum of the group sums for each row, or level, of Factor A; and there is an extra row at the bottom which gives the sum of the group sums for each column, or level, of Factor B.* The entry in the lower right hand corner is *the grand sum* of all *abn* cases, $\Sigma_1^{abn} X$. This quantity, we have just seen, appears in the last term of both the formula for SS_T and formula *12.2* for SS_B.

Calculation of *SS*'s for the Components of SS_B

We shall now repeat the calculation of SS_B by formula *12.2* in order to bring out the parallel procedures involved in the calculation of its component parts, *SS* for *A* and *SS* for *B*. (*Parts in brackets not needed for calculation.*)

$$SS_B = \left[n \sum_1^{ab} (\bar{X}_{ab} - \bar{X}_{abn})^2 \right]$$

$$= \sum_1^{ab} \frac{\left(\sum_1^n X_{ab}\right)^2}{n} - \frac{\left(\sum_1^{abn} X\right)^2}{abn} \quad (12.2)$$

$$= \frac{(20)^2 + (30)^2 + (40)^2 + (10)^2}{5} - \frac{(100)^2}{2 \times 2 \times 5}$$

$$= 600 - 500 = 100$$

$$\frac{SS \text{ for } A}{(\text{Sex})} = \left[bn \sum_1^a (\bar{X}_a - \bar{X}_{abn})^2 \right]$$

$$= \sum_1^a \frac{(\Sigma X_a)^2}{bn} - \frac{\left(\sum_1^{abn} X\right)^2}{abn} \quad (12.4)$$

$$= \frac{(50)^2 + (50)^2}{2 \times 5} - 500 = 500 - 500 = 0$$

$$\frac{SS \text{ for } B}{(\text{Texts})} = \left[an \sum_{1}^{b} (\bar{X}_b - \bar{X}_{abn})^2 \right]$$

$$= \sum_{1}^{b} \frac{(\Sigma X_b)^2}{an} - \frac{\left(\sum_{1}^{abn} X\right)^2}{abn} \quad (12.5)$$

$$= \frac{(60)^2 + (40)^2}{2 \times 5} - 500$$

$$= 520 - 500 = 20$$

$$SS_{A \times B} = SS_B - SS \text{ for } A - SS \text{ for } B$$
(Interaction)
$$= 100 - 0 - 20 = 80 \quad (12.6)$$

Comment on Formulas

Before summarizing our findings from these calculations, let us look at the formulas involved for a minute. The first thing we observe is that *the last terms of formulas 12.2, 12.4 and 12.5 are identical* and this last term is also identical with the last term of the formula for SS_T, 12.1. In order to achieve this great saving of calculation time we did some algebraical acrobatics, which left the parts of the formulas we used (the parts not in brackets) showing little resemblance to *the parts in brackets which we do not have to calculate*. Though in the calculations we worked entirely with *sums*, the parts in brackets involve *means* and tell us what is really going on. *In the formulas for SS_B and its components we are primarily concerned with the deviation of various means from the grand mean of all abn cases.*

If we look at the parts in brackets in these three formulas we see that $\bar{X}_{abn}$, the grand mean, appears in all of them. The basic ingredient in the SS_B formula is $(\bar{X}_{ab} - \bar{X}_{abn})$, the deviation of the group means $(\bar{X}_{ab})$ from the grand mean. The sum of the squared deviations is multiplied by n because each group mean is based on n cases. In the SS for A formula the basic ingredient is $(\bar{X}_a - \bar{X}_{abn})$. $\bar{X}_a$ represents the mean for each of the various rows, or levels, of Factor A, as a takes values $1, 2, \cdots a$. (In this case there are only two rows or levels.) In this formula the sum of the squared deviations of *the level means* from the grand mean is multiplied by bn, because on each row there are b groups with n cases in each. The formula for SS for B is made up in the same way but it applies to the columns, or levels, of Factor B, instead of to the rows, or levels, of Factor A.

Alternative Method for Calculating SS's

It should be possible to calculate these three SS's directly from the bracket formulas. To do this we convert all the sums in Table 16, both the group sums and the marginal, or level sums, into means, as in Table 17.

TABLE 17

Group Means and Level Means for a 2 × 2 Factorial Experiment Based on the Sums in Table 16
($n = 5$)

Sex	Textbooks		Row Means
	B_1 (new)	B_2 (old)	$\bar{X}_a$ (for $bn = 10$)
A_1 (boys)	4.0	6.0	5.0
A_2 (girls)	8.0	2.0	5.0
Column Means $\bar{X}$ (for $an = 10$)	6.0	4.0	$5.0 = \bar{X}_{abn}$

Using the "bracket formulas":

$$SS_B = n \sum_1^{ab} (\bar{X}_{ab} - \bar{X}_{abn})^2$$
$$= 5[(-1)^2 + 1^2 + 3^2 + (-3)^2]$$
$$= 5 \times 20 = 100 \qquad \text{(From 12.2)}$$

$$SS \text{ for } A = bn \sum_1^a (\bar{X}_a - \bar{X}_{abn})^2$$
$$= 10(0^2 + 0^2) = 0 \qquad \text{(From 12.4)}$$

$$SS \text{ for } B = an \sum_1^b (\bar{X}_b - \bar{X}_{abn})^2$$
$$= 10[1^2 + (-1)^2] = 20 \qquad \text{(From 12.5)}$$

All these SS values check with those obtained by the other part of the same formulas.

The procedure here seems ridiculously simple! Why not use it all the time? The reason is that the data in Tables 16 and 17 were artificially simplified, in order to reduce the arithmetical labor and thus to make the general procedures more intelligible. However, when the means in the bracket formulas involve decimals, the calculation is more difficult. Furthermore, when we work with sums, rather than means, a smaller number of operations is usually involved, though the numbers may be larger. And *with a calculating machine the sum formulas are much easier to work with*, because the machine can get the sum of a series of numbers and the sum of their squares in the same general operation. Therefore, though the formulas expressed in terms of means are more meaningful (sorry!), it is usually preferable to do our calculating with the formulas expressed in terms of sums, if possible.

Calculation of F Ratios

Now let us return to the SS's in our experiment (obtained by either method) and complete the analysis. *The appropriate d.f.'s for the calculation of the F ratios and for use in their interpretation* are included in the following summary.

Summary of Stage II: Analysis of Variation between Groups into Main Effect for A (Sex), Main Effect for B (Textbooks), and Interaction Effect

Source	SS	d.f.	MS	F^*
Components of SS_B:				
A (Sex)	0	$(a-1) = 1$	0	$0/2.5 = 0$
B (Texts)	20	$(b-1) = 1$	20	$20/2.5 = 8$
A × B (Interaction)	80	$(a-1)(b-1) = 1$	80	$80/2.5 = 32$
From Preliminary Analysis:				
Between Groups	100	$(ab-1) = 3$	33.3	$33.3/2.5 = 13.3$
Within Groups	40	$ab(n-1) = 16$	2.5	
Total	140	$abn - 1 = 19$		

Note: For every case in this table the denominator of the F ratio is MS_W obtained from the preliminary analysis.

Interpretation of F Ratios

Now how do we interpret the F ratios? Since for Factor A $F = 0$, we can say without hesitation (almost) that the main effect of sex is not significant.* It might be more diplomatic to put it this way: the improvement in arithmetic of *the average* fifth-grade boy (combining the samples from both textbook groups) does not differ significantly from that of *the average* fifth-grade girl (from both book groups) under the conditions of our experiment.

But for the textbook factor (B) we had $F = 8$. Entering the table of F in the row for $d.f. = 1$, we find in the column for $d.f. = 16$ that $F = 4.49$ for $p = .05$ and 8.53 for $p = .01$. Therefore, the main effect of the textbook factor is significant at the $p < .05$ level. Put another way, we can say that *the average* improvement of the boys and girls *taken together* on the new arithmetic book is significantly greater than their *average* performance on the old book.

The F ratio for the interaction effect was 32. Since this is much greater than 8.53, the tabled value of F for $p = .01$ for the proper $d.f.$'s of 1 and 16, we have a highly significant interaction effect. We can get an idea of *the nature of this interaction effect* if we look at the 2×2 table of means for our experiment (p. 141). Though the new textbook led to better results than the old one *on the average* (as indicated in the column means), the effects were not clear and unequivocal. The boys actually did less well with the new book. It was only because the girls did very much better with it that the combined average for boys and girls was better. Thus, we can see that the two factors, sex and textbook, are not independent of each other, but they "interact."

Now what decision can be made on the basis of these findings? Before deciding *it might be well to follow up the variance*

---

*In the *general* discussion of F distributions (section 2) we learned that the values of F close to zero are almost as unusual as very large values (and might be the basis for rejecting some null hypothesis). However, in most analyses, including the present one, we are interested only in evidence that the variance (MS) between groups, or levels, etc., is significantly greater (not less) than the variance within groups (MS_W). In other words, we are using a *one-tailed significance test*. And Table 13 is intended for this purpose.

analysis with a Scheffé test on certain specific comparisons. Of the six possible comparisons of two means there are only two which are relevant to the purpose of our experiment: boys-new vs. boys-old and girls-new vs. girls-old; that is, $\bar{X}_{11} - \bar{X}_{12}$ and $\bar{X}_{21} - \bar{X}_{22}$. Since there are only two means involved in these comparisons, we can use formula *11.9*:

$$\text{Min. Diff. for Sig.} = t' s_{\text{Diff.}}$$

where

$$s_{\text{Diff.}} = \sqrt{\frac{2MS_W}{n}} = \sqrt{\frac{2 \times 2.5}{5}} = 1.0$$

and

$$t'_{.05} = \sqrt{(k-1)F'} = \sqrt{3 \times 3.24} = 3.12$$

(where F' is the tabled value of F for 3 and 16 *d.f.* and $p = .05$)

and

$$t'_{.01} = \sqrt{3 \times 5.29} = 3.98$$

This gives us

$$\text{Min. Diff. for Sig.}_{.05} = 3.12$$

and

$$\text{Min. Diff. for Sig.}_{.01} = 3.98$$

Therefore, since $\bar{X}_{11} - \bar{X}_{12} = 4.0 - 6.0 = -2.0$, and the absolute value of this difference is $2.0 < 3.12$, the boys do not do significantly worse with the new book. But since $\bar{X}_{21} - \bar{X}_{22} = 8.0 - 2.0 = 6.0 > 3.98$, the girls do significantly better with the new book at the $p < .01$ level.

With this additional information on the significance of the observed changes in performance, the school might decide that the new book was a fair risk. Or the decision might be to make no change until after another experiment with larger samples.

It is only fair to observe that we might have come close to these same conclusions from an examination of the table of means, supplemented by two *regular t* tests, without using the variance analysis at all. (We could not use the Scheffé *t* test, of course, without first calculating MS_W, which is part of the variance analysis.) But let us remember that we are working with a very simple experimental design, in order to illustrate the analysis procedures. If we had used three or more levels for each experimental factor, the more elaborate analysis would have been clearly justified. Let us now run through the main steps again quickly, without any detailed explanations, for a two-factor experiment with three levels for each factor, a 3 × 3 experimental design.

3. Two-Factor, Three-Level Experiment on Drug Effects

The drugs Thorazine and Stelazine are known to be very helpful in the treatment of certain kinds of mental disorder. In this final illustration of analysis of variance procedures we shall assume data from *an experiment designed to throw light on the optimum dosages of these two drugs in combination.* This calls for a two-factor design, with our attention focused on the *interaction effect*, if any. We shall use three levels (or dosages) of each drug. The first level of each is no drug at all. This gives us *a control group*, which provides a reference point, a measure of the patients' response to the general hospital routine without any drug therapy. The dosage for the third level of each drug is twice that for the second level.

In order to avoid repetition of the more obvious steps in the preliminary analysis (Stage I), we shall start with a 3 × 3 table of *the sums* of the X's, the "improvement scores," for the $n = 10$ patients in each of the nine treatment groups, groups involving all possible combinations of the three levels of each of the two drug factors. We shall work with the treatment sums rather than the individual scores of all the patients. Table 18 really gives us all the information we need for a complete variance analysis except $\Sigma_1^{abn} X^2$, the sum of the X^2's for all $abn = 90$ cases. *We shall assume that the value for this, 1172.5, has been worked out earlier* from the complete data. *The symbols in this table have been reduced to the simplest possible terms.* The equivalents of these new symbols which we used earlier are given in the lower part of the table.

TABLE 18

Sums for the Nine Factor Combinations for a 3×3 Factorial Experiment on Two Drugs
($n = 10$ for each cell)

		\multicolumn{3}{c}{Levels of Factor B (Stelazine)}	
		B_1	B_2	B_3	ΣR (for $bn = 30$)
Levels of	A_1	10	20	30	60
Factor A	A_2	25	45	20	90
(Thorazine)	A_3	40	55	25	120
ΣC (for $an = 30$)		75	120	75	$270 = GS = \sum_1^{abn} X$

The simplified symbols used with this table and in the SS equations below are:

$$CE = \text{Cell Entry} = \text{treatment sum} = \sum_1^n X_{ab}$$

$$\Sigma R = \Sigma \text{ Rows} = \text{level sums for } A = \Sigma X_a$$

$$\Sigma C = \Sigma \text{ Columns} = \text{level sums for } B = \Sigma X_b$$

$$GS = \text{Grand Sum} = \sum_1^{abn} X$$

The SS Calculations Based on the Simplified Notation of Table 18

$$SS_T = \sum_1^{abn} X^2 - \frac{(GS)^2}{abn} = \underset{\text{(given)}}{1172.5} - \frac{(270)^2}{3 \times 3 \times 10}$$
$$= 1172.5 - 810 = 362.5 \qquad (12.7)$$
$$\text{(modified } 12.1\text{)}$$

$$SS_B = \sum_1^{ab} \frac{(CE)^2}{n} - \frac{(GS)^2}{abn} \qquad \begin{array}{c}(12.8)\\ \text{(modified } 12.2\text{)}\end{array}$$
$$= \frac{(10)^2 + (20)^2 + (30)^2 \ldots (55)^2 + (25)^2}{10} - 810$$
$$= 970 - 810 = 160$$

$$SS \text{ for } A = \sum_1^a \frac{(\Sigma R)^2}{bn} - \frac{(GS)^2}{abn} \qquad (12.9)$$
$$\text{(modified } 12.4)$$
$$= \frac{(60)^2 + (90)^2 + (120)^2}{3 \times 10} - 810$$
$$= 870 - 810 = 60$$

$$SS \text{ for } B = \sum_1^b \frac{(\Sigma C)^2}{an} - \frac{(GS)^2}{abn} \qquad (12.10)$$
$$\text{(modified } 12.5)$$
$$= \frac{(75)^2 + (120)^2 + (75)^2}{3 \times 10} - 810$$
$$= 855 - 810 = 45$$

$$SS_{A \times B} = SS_B - SS \text{ for } A - SS \text{ for } B$$
$$= 160 - 60 - 45 = 55 \qquad (12.6)$$
$$SS_W = SS_T - SS_B = 362.5 - 160 = 202.5 \qquad (12.3)$$

The Calculation of the F Ratios

The SS results are summarized and the F ratios are calculated in the table on the next page. This table also gives *the proper d.f.'s for the evaluation of the F ratios.*

Interpretation of the F Ratios

From the preliminary analysis we find from Table 13 that for 8 and 81 $d.f.$ $F = 2.74$ for $p = .01$. (The table does not list values for 81 $d.f.$; but the listed values for 80 $d.f.$ do not differ appreciably even in the second decimal place.) Since our obtained value of $8.00 > 2.74$, we may conclude that MS_B is significantly greater than MS_W at the $p < .01$ level, or that *there is a highly significant overall treatment effect.*

All of the component variances are also significant. This is clear from the tabled values of F. For 2 and 81 $d.f.$ $F = 4.88$ for $p = .01$. Since $12.00 > 4.88$, *the main effect of Thorazine is highly significant* at the $p < .01$ level. And since $9.00 > 4.88$, *the main effect of Stelazine is also highly significant* at the $p < .01$ level. For the interaction effect we must use 4 and 81 $d.f.$ For this the tabled value of F is 3.56 for $p = .01$. Since $5.50 > 3.56$, *there is also a highly significant interaction effect* at the $p < .01$ level.

Two-Factor Analysis of Variance

Summary of the Variance Analysis for a Two-Factor Three-Level Experiment on Drugs

Source	SS	d.f.	MS	F
Components of SS_B				
A (Thorazine)	60	$(a-1) = 2$	$60/2 = 30.0$	$30.0/2.5 = 12.00$
B (Stelazine)	45	$(b-1) = 2$	$45/2 = 22.5$	$22.5/2.5 = 9.00$
A × B (Interaction)	55	$(a-1)(b-1) = 4$	$55/4 = 13.75$	$13.75/2.5 = 5.50$
	(160)	(8)		
Preliminary Analysis				
Between	160	$(ab-1) = 8$	$160/8 = 20.0$	$20.0/2.5 = 8.00$
Within	202.5	$ab(n-1) = 81$	$202.5/81 = 2.5$	
Total	362.5	$(abn-1) = 89$		

Note: For every case in this table the denominator of the F ratio is MS_W obtained from the preliminary analysis, 2.5.

Scheffé Comparisons

Having found that there are significant main effects for both drugs, and also a significant interaction effect, let us see if we can locate some useful treatment effects more precisely. It will help if we convert our table of treatment sums into a table of treatment means. Since $n = 10$, this can be done by simply inserting a decimal point after the first digit in the sums. Table 19 gives us the results. Before we proceed to make any specific

TABLE 19

Treatment Means for a 3 × 3 Factorial Experiment on Two Drugs
($n = 10$)

		Levels of Factor B (Stelazine)		
		B_1	B_2	B_3
Levels of	A_1	1.0	2.0	3.0
Factor A	A_2	2.5	4.5	2.0
(Thorazine)	A_3	4.0	5.5	2.5

tests, we can get some impressions from this table as to the effect of each drug by itself and some impressions of how the drugs act in combination with each other.

If we look at the first row, or level A_1 which involves no Thorazine dosage at all, we observe an apparent improvement (from 1.0 to 3.0) as the Stelazine dosage increases. In the first column, level B_1 which involves no Stelazine, we observe an apparent improvement (from 1.0 to 4.0) as the Thorazine dosage is increased. But *we get the most marked improvement when the two drugs are combined.* When we compare the control group (A_1B_1), which has no drug at all, with group A_2B_2, which has moderate dosages of both drugs, we get an apparent improvement of from 1.0 to 4.5. And the comparison $A_3B_2 - A_1B_1$, which involves the large dose of Thorazine and the moderate dose of Stelazine in contrast with the control group, brings out the largest apparent improvement, from 1.0 to 5.5. We also observe that the mean improvement scores drop off when we use the large dose of Stelazine in combination with either dosage of Thorazine. *Interaction can thus be either positive or negative.*

Now let us check our impressions by using specific Scheffé tests of significance. We can easily work out the value of *the minimum difference between any two means that will be necessary for significance* at either the $p = .05$ or $p = .01$ levels. The formula is

$$\text{Min. Diff. for Sig.} = t's_{\text{Diff.}} \qquad (11.9)$$

where

$$s_{\text{Diff.}} = \sqrt{\frac{2MS_W}{n}} = \sqrt{\frac{2 \times 2.5}{10}} = .707$$

and

$$t'_{.05} = \sqrt{(ab - 1)F'} = \sqrt{8 \times 2.05} = 4.05$$

(where F' is the tabled value of F for 8 and 81 d.f. and $p = .05$)

and

$$t'_{.01} = \sqrt{8 \times 2.74} = 4.68$$

(where 2.74 is the tabled value of F for 8 and 81 d.f. and $p = .01$)

This gives us

$$\text{Min. Diff. for Sig.}_{.05} = 4.05 \times .707 = 2.86$$

and

$$\text{Min. Diff. for Sig.}_{.01} = 4.68 \times .707 = 3.31$$

We can now quickly test the four comparisons which our inspection of the table of means suggested.

(C_1) $\bar{X}_{13} - \bar{X}_{11} = 3.0 - 1.0 = 2.0 < 2.86$

(C_2) $\bar{X}_{31} - \bar{X}_{11} = 4.0 - 1.0 = 3.0 > 2.86$

(C_3) $\bar{X}_{22} - \bar{X}_{11} = 4.5 - 1.0 = 3.5 > 3.31$

(C_4) $\bar{X}_{32} - \bar{X}_{11} = 5.5 - 1.0 = 4.5 > 3.31$

We conclude from C_1 that the large Stelazine dose by itself does not produce a significant improvement; but from C_2 we conclude that the large Thorazine dose does at the $p < .05$ level. From comparisons C_3 and C_4 we conclude that the moderate dose of Stalazine, in combination with either the moderate or the large doses of Thorazine, produces a highly significant difference in the mean improvement scores of the patients at the $p < .01$ level.

In practice, of course, it is desirable to keep doses down to the lowest levels at which they are effective, in order to reduce or eliminate side effects. Also, in practice *individuals differ widely in their response to drugs*. But an experiment similar to the one we have just analyzed, using somewhat artificial data, could set up useful guide lines for practical treatment.

4. Fixed Effects and Other Experimental Designs*

The experimental designs used in this and the preceding chapter (with both one and two factors) are examples of the so-called "fixed effects" design. Though the discussion in these two chapters applies only to this general type of design, the basic reasoning may be extended to other designs, such as the "random effects" and "mixed" designs, which may be more elaborate.

The general distinction between these three types of design may be brought out by the drug study above (Tables 18 and 19). As described and analyzed, the study is an illustration of *the "fixed effects" design*. In it we used, in addition to the control (or zero) dosages, one "moderate" and one "large" dosage only of each of the two drugs. Though any number of possible dosages might have been used, the experimental interest in this case was restricted to the effect of the various combinations of the two specific dosages actually used. Hence, the inferences drawn from the analysis of the experimental results applied only to the particular treatment combinations employed. The inferences were relevant or "fixed" to these combinations alone.

Now it is obvious that there might be an interest in the two drugs broader than in the effects of just the two levels studied. More information could of course be obtained by increasing the number of levels for one or both of the drugs. This would still be an example of the fixed effects design. But if, from

---

*This section may be omitted at the discretion of the instructor.

a very large number of possible safe dosage levels, we take *a random sample* and then use these sample levels in our experiment, we have the basis for *the "random effects" design*. This level-sampling can be used with either one drug or two, one factor or two factors. The calculation procedures will be much the same as for the fixed effects design; but the hypotheses and inferences will be different. In the random effects design we make inferences about the whole range of possible levels on the basis of the sample of levels actually used in the experiment. In *the "mixed" experimental design* one drug, or other factor, has fixed levels while the other factor levels (or the levels of other factors, if more than two factors are involved) are chosen by random sampling. The details of these elaborate designs are beyond the scope of this short text. However, what we have learned about fixed effects designs is not only useful in itself but is the basis for an understanding of more complex designs.*

Exercises

1. The four sets of scores below on a certain attitude scale are random samples from four different populations: men and women psychology majors and men and women education majors. The experimental interest is in a possible difference in attitudes between: (1) men and women in general and (2) psychology majors and education majors, regardless of their sex. (3) There is also interest in a possible "interaction" effect between the two factors, the sex of the subjects and their major field of study.

Men (A_1)		*Women* (A_2)	
Psychology (B_1)	*Education* (B_2)	*Psychology* (B_1)	*Education* (B_2)
X_{11}	X_{12}	X_{21}	X_{22}
−2	8	6	3
0	5	4	5
1	7	3	2
2	6	7	6
4	4	5	4

*For further details and a more advanced treatment of variance analysis see W. L. Hays, *Statistics for Psychologists*. Holt, Rinehart and Winston, 1963, chapters 12 and 13.

(a) For each set get the "treatment" sum, the mean, and ΣX^2.
(b) Start the preliminary variance analysis by getting SS_T, SS_B and SS_W. (c) Calculate and interpret F, giving H_o and conclusion.

2. Set up the "treatment" *sums* for the data of Exercise 1 in a 2 × 2 table and indicate the two *level* sums in the margins for each of the two factors. (a) Get SS for A, SS for B, and the interaction SS. (b) Set up the corresponding table of "treatment" *means* and level means and check SS_B, SS for A, and SS for B by the "bracket formulas" of p. 141. (c) Find the F ratios for the main effects of sex and major field and for the interaction effect. (d) Interpret the F's, giving for each the level of significance, if any.

3. Apply the Scheffé significance test to the six possible differences between two means for the four groups in Exercise 1, using the following steps: (a) Find the common $s_{\text{Diff.}}$. (b) Find $t'_{.05}$ and $t'_{.01}$. (c) Find $Min.\ Diff.\ for\ Sig.$ for $p = .05$ and $p = .01$. (d) What differences, if any, are significant and at what level?

4. Given the following treatment sums for variable X for a 2 × 4 factorial experiment, with $n = 10$ for each cell.

		Levels of B			
		1	2	3	4
Levels of A	1	5	10	20	25
	2	10	20	40	30

Also given: $\Sigma_1^{abn} X^2 = 559$. Using the simplified procedure of pp. 146f. in a preliminary analysis: (a) Find SS_T, SS_B and SS_W. (b) Calculate and interpret F, giving H_o and conclusion.

5. Get the row and column (or level) sums for the data of Exercise 4. Then: (a) Find SS for A, SS for B, and $SS_{A \times B}$. (b) Find the F ratios for the main effects of A and B and for the interaction effect. (c) Interpret the F's, giving level of significance, if any.

6. Convert the treatment sums of Exercise 4 into a 2 × 4 table of treatment means. Then apply the Scheffé test to the 28 possible differences between two means, using steps (a) to (d) as in Exercise 3.

13
Correlation Techniques; Reliability of Tests and Confidence Limits for Test Scores

In this chapter we shall deal with techniques for investigating relationships between two sets of scores from the same sample of subjects. It is often desirable, for example, to have some way of indicating how performance on one test (say an intelligence test) is related to performance on another test (say a final examination). We might like to know whether, in general, those who do well on one test also do well on the other, or whether there is no relationship whatsoever. Or again, in other fields, we might like to have some means of indicating the degree of parallelism between measures of height and weight, or between rainfall and crop yield, or between slum congestion and delinquency rates, and so on. The degree of parallelism, the extent to which two sets of measures vary in unison, is commonly expressed by means of a correlation coefficient. Two common methods for determining correlation coefficients will be discussed below.

1. The Rank-Difference Correlation Method

The simpler method for obtaining a correlation coefficient is the rank-difference method. It is best suited to studies involving a small number of cases, say 15 to 30. The procedure is illustrated in Table 20, with simplified data. We have here the measures of height and weight for 10 subjects in parallel columns. The first step is to rank the measures of height, assigning rank 1 to the largest and rank n (in this case 10) to the smallest, and intermediate ranks to the scores between these extremes. The *tied scores* are the only ones to cause any confusion; but these are handled in logical manner. For example, the two 71's, had they not been tied, would have filled rank positions 2 and 3; hence we split the points between them and assign the average rank, 2.5, to each. In the case of the triple tie between the 68's for ranks 6, 7, and 8, each is assigned the average rank again—in this case, 7. But we must be careful in ranking the next measure to skip rank 8, which has already been assigned. In similar fashion we rank the other set of measures, the weights. Then, as the name of the method suggests, we fill in the D column with the differences in ranks, ignoring the minus signs. We finally square each D value, obtain ΣD^2, and substitute in the formula.

The symbol r' is here used to represent the rank-difference correlation coefficient. It is used to suggest similarity to, but not identity with, the product-moment coefficient discussed below, the symbol for which is r. The notation r'_{HW} is a convenient shorthand for "r' between height and weight." In this illustration we get a positive correlation between height and weight, which is what we should expect. However, there is not a one-to-one correspondence (i.e., tallest man, the heaviest; second tallest, second heaviest; and so on). If there had been a perfect correlation, r' would have been equal to 1.00. A perfect inverse relationship would have given a correlation of -1.00. No relationship at all would have been indicated by a correlation of .00.

156 Correlation Techniques

TABLE 20
Rank-Difference Correlation Technique Illustrated
(SKELETON DATA)

Subject	Height (Inches)	Weight (Pounds)	R_H (Ranking in Height)	R_W (Ranking in Weight)	D ($R_H - R_W$)	D^2
1	70	165	4	3	1	1
2	66	130	9	10	1	1
3	72	180	1	1	0	0
4	68	145	7	7.5	0.5	0.25
5	71	160	2.5	4.5	2	4
6	64	150	10	6	4	16
7	68	140	7	9	2	4
8	71	168	2.5	2	0.5	0.25
9	69	145	5	7.5	2.5	6.25
10	68	160	7	4.5	2.5	6.25

$n = 10$ (Ignore signs) $\Sigma D^2 = 39.0$

$$r' = 1 - \frac{6 \Sigma D^2}{n(n^2-1)} ; r'_{HW} = 1 - \frac{6 \times 39.0}{10 \times 99} = .76 \qquad (18.1)$$

In the illustration in the table only 10 cases were used, in order to make the procedure clearer, but in practice a correlation based on less than 15 or 20 cases would not be taken very seriously. Though the rank-difference method is simpler when the number of cases is small and is sufficiently reliable for purposes of estimating the correlation, it should be replaced by the more elaborate product-moment method in more refined investigations, *or when there are many tied scores.*

2. The Product-Moment Method of Correlation

The Pearson product-moment correlation index, r, is more reliable than r', for it takes into account the absolute size of the measures and not merely their rank-order. The method of obtaining r is illustrated with skeleton data, ungrouped, in Table 21.

In this table r stands for the product-moment correlation coefficient between two sets of measures, X and Y; x and y stand for the deviations from the means, $\bar{X}$ and $\bar{Y}$, respectively. In correlation problems n stands for the number of *pairs* of scores used. In testing *the significance of r* a special formula for t may be used:

$$t = \frac{r\sqrt{n-2}}{\sqrt{1-r^2}} \tag{13.3}$$

For large values of n this t may be interpreted with a fair degree of accuracy from Table 10. For values of n of 32 or less Table 11 gives us more accurate results. In using Table 11 for our present purpose we use $d.f. = n - 2$. No matter which table we use, we are in effect testing the *null hypothesis* that our sample of pairs of X and Y scores are drawn at random (by chance) from a larger population in which the correlation between the X and Y variables is zero. If the tables tell us that an obtained value of t is very improbable on the basis of this hypothesis, we reject it and conclude that something other than chance was responsible for the obtained value of r. In other words, we conclude that r is significantly different from zero (at whatever p level the tables indicate) and that there really is some relation between X and Y.

TABLE 21

Product-Moment Correlation Technique Illustrated
(SKELETON DATA, UNGROUPED)

S	Measures		x	y	x^2	y^2	xy	
	X	Y					(+)	(−)
1	19	15	4	5	16	25	20	
2	16	11	1	1	1	1	1	
3	15	10	0	0	0	0		
4	11	7	−4	−3	16	9	12	
5	17	8	2	−2	4	4		−4
6	14	9	−1	−1	1	1	1	
7	18	13	3	3	9	9	9	
8	13	12	−2	2	4	4		−4
9	12	5	−3	−5	9	25	15	
							(58)	(−8)

$n = 9$ $\bar{X} = 15$ $\bar{Y} = 10$ $\Sigma x^2 = 60$ $\Sigma y^2 = 78$ $\Sigma xy = 50$ (algebraic sum)

$$r = \frac{\Sigma xy}{\sqrt{\Sigma x^2 \cdot \Sigma y^2}} = \frac{50}{\sqrt{60 \times 78}} = .73 \qquad (13.2)$$

$$t = \frac{r\sqrt{n-2}}{\sqrt{1-r^2}} \qquad (13.3)$$

In this case $t = \dfrac{.73\sqrt{9-2}}{\sqrt{1-(.73)^2}} = \dfrac{.73 \times 2.65}{\sqrt{1-.533}} = 2.83$

Using Table 11 and $d.f. = n - 2$, we find that r is significantly different from zero at something not far from the $p = .02$ level, certainly at the $p < .05$ level.

For example, in our illustrative problem in Table 21 an r of .73 for 9 pairs of scores gave us a t of 2.83. Looking this up in Table 11 for $d.f. = 7$ (that is, $9 - 2$), we find that it lies between the t values for $p = .05$ and $p = .02$, but much closer to the value for $p = .02$. We can then say that our correlation was significantly different from zero at about the $p = .02$ level. (This test, unfortunately, is appropriate for use with the rank-difference correlation only when there are no tied ranks and when $n > 10$.) If a correlation proves to be significant, there then arises the question of its interpretation. This question is taken up in the next section.

3. Interpretation of Correlation Coefficients

First and foremost, in the interpretation of these coefficients, we should keep in mind that even a perfect correlation between two variables, X and Y, does not indicate any necessary or specific *causal* relationship between them; it merely implies that X and Y vary in perfect unison, if $r = 1.00$, and in perfect opposition, if $r = -1.00$. Logically, variations in X might be caused by variations in Y, or variations in Y might be caused by variations in X, or variations in both X and Y might be caused by variations in some third variable, Z.

Secondly, in interpreting coefficients we must start out by checking on their significance, as suggested above. If they are not significantly different from zero, the question of their interpretation does not arise; there is just nothing to interpret. But, if they do differ significantly from zero, the interpretation depends upon the purpose for which they have been calculated. Our purpose is sometimes to *estimate* scores on one of two correlated tests from a knowledge of scores on the other. Sometimes we wish to *predict* probable standing in one of two correlated distributions (say college grades) from standing in the other (say high school grades or intelligence test scores). For either of these purposes coefficients must be quite high to be useful.

Let us suppose, for example, that we wish to estimate John's score in test Y from a knowledge of his score in test X. If r were zero, a knowledge of his score in X would not help us at all in estimating his score in Y. The best estimate we could make under these circumstances would be the mean of test Y; and the *standard error of* such an *estimate* would be the standard deviation (s) of the Y distribution.* But if r were equal to 1.00 (or to

---

*This is just another way of saying that the chances are about 68 in 100 that John's score will fall between the limits $\overline{Y} - 1s$ and $\overline{Y} + 1s$, etc., if the distribution of Y is normal. But we know this already without any knowledge of his X score. In section 3 of the next chapter the standard error of estimate is discussed more fully.

−1.00), we could estimate his score in Y with perfect accuracy by means of a *regression equation* (see next chapter), and the standard error of such an estimate would be zero. Thus, as r changes from .00 to 1.00 (or −1.00), the error of estimate is reduced from a finite value to zero. But this reduction in the error of estimate does not bear any simple linear relation to r. An r of .50, for example, does *not* reduce the error of the estimate 50%, but only about 13%. To reduce the error of the estimate 50% we must have an r of approximately .87. This being the case, estimates of *individual* scores based on r's smaller than .50 are of limited value. This is discussed in more detail in the next chapter.

More often, however, our purpose is not to estimate or to predict individual scores, but our interest is in *general* tendencies, or in the *relative* closeness of relationship indicated by two or more correlation coefficients. In this field smaller values of r may be useful, but here also we must ordinarily guard against a simple linear interpretation of the coefficients. They cannot be thought of as percentages, for example. Thus, we cannot say that an r of .40 is just one-half as good as an r of .80. However, for many purposes, we can get an *approximate* idea of the *relative* closeness of relationship indicated by two r's if we compare their squares. That is, an r of .40 is only about one-fourth as good as an r of .80. It is impossible to give briefly and in familiar terms a more explicit interpretation of these coefficients, but a little more light on the subject will appear in connection with certain applications discussed below and in the next chapter.

4. Reliability and Validity of Tests

Checking the reliability of tests and checking their validity are two important applications of the correlation technique. When we ask, "How reliable is that test?" we mean, "How self-consistent is it?" We want to know how stable are the results obtained from it; we want to know whether, in general, those who get good scores on one occasion will get good scores on a second try at it. Obviously, if the rank-order of scores obtained in a second administration of a test differs markedly from the order obtained in the first, the test is worthless as an index of individual ability. Hence, a fair degree of reliability is a minimum requirement for any test. The *reliability of a test* is commonly estimated by correlating it with itself in one of the three following ways: test with retest, one form with another equivalent form, or one half with the other. The values derived from these correlations are called *reliability coefficients* and are designated by the symbol r_{11}. The last of these three methods, sometimes called the *split-half technique*, is usually carried out by correlating odd-numbered items with even-numbered items. The procedure gives the reliability for a test which is only half its original length. Since the usual effect of shortening a test is to lessen its reliability, it is customary to make allowance for this and to estimate the full-length reliability by means of the Spearman-Brown formula below:

$$r_{11} = \frac{2r_{1/2}}{1 + r_{1/2}} \qquad (13.4)$$

where r_{11} is the estimated reliability coefficient for the whole test and $r_{1/2}$ is the value obtained by correlating the two halves. For example, if $r_{1/2}$ is found to be .50, then

$$r_{11} \text{ will equal } \frac{2 \times .50}{1 + .50} = \frac{1.00}{1.50} = .67$$

A test which is to be used to differentiate between individuals should have a reliability of .90 or better; but, if a test is merely intended to differentiate between the means of two or more groups, a reliability in the neighborhood of .80 is adequate.

There is another aspect of a test which must be checked before a test can be used intelligently—its *validity*. When we ask, "How valid is that test?" we mean, "How closely does the test measure what it purports to measure?" In attempting to answer this question we again make use of the correlation technique, but validity is not as easy to pin down as reliability. The difficulty lies in finding a suitable *criterion* of what the test is trying to measure, outside of the test itself, against which the test may be checked by correlation. For example, in the case of an intelligence test, what shall we take as a true measure of intelligence? Various criteria have been proposed, such as grades in school, the combined ratings of several teachers or other judges, another test already validated, and so on. There is no general rule for the selection of validation criteria; they will vary from test to test.* However, this problem may be left to the individual test maker; our primary concern here is with the statistical techniques themselves. It is customary to consider a test valid if a high correlation is found between the test and a satisfactory independent criterion, provided the reliabilities of both the test and the criterion are satisfactory. We do not ordinarily expect, however, that a correlation between a test and its criterion will be as high as a self-correlation, or reliability coefficient.

5. Confidence Limits for Individual Test Scores

It is possible, with the aid of the reliability coefficient, to set up confidence limits for individual test scores. Individual scores, like all other measures, are subject to chance variations from their theoretically true values. The effect of such variations, or "chance errors," may be estimated from the formula

$$s_{1\infty} = s_1\sqrt{1 - r_{11}} \qquad (13.5)$$

where $s_{1\infty}$ stands for the *standard error of an obtained score* (or the *standard error of measurement*), s_1 for the standard deviation of the test in question, and r_{11} for its reliability coefficient.† To

---

*Sometimes, when no external criteria are available for validating a test, the *"criterion of internal consistency"* is used: a large number of tentative test items are administered in a preliminary test; then, only those items are used in the final selection which distinguished clearly between the subjects with high scores and those with low scores, usually in terms of the percentage of subjects in each group which gave a particular kind of response.

†If the reliability coefficient was obtained by correlating one form of the test with another (rather than by the retest or split-half techniques), we may replace s_1 in the above formula with the average of the s's for the two forms.

illustrate the use of this formula, let us assume that $s_1 = 5$ and $r_{1I} = .84$. Then $s_{1\infty} = 5\sqrt{1 - .84} = 5 \times .4 = 2.0$. We interpret the standard error of measurement for any obtained score, X, in terms of the confidence limits discussed in Chapter 7. If we repeatedly assert that the true score lies between the limits $X - 1s_{1\infty}$ and $X + 1s_{1\infty}$, on the basis of a great many sample scores from the same population, we should expect to be right 68 times out of 100. Or, if we repeatedly set the confidence limits for the true score between $X - 2.58s_{1\infty}$ and $X + 2.58s_{1\infty}$, we should expect to be right nearly 100 times out of 100. If we take as an obtained score 60, say, in the concrete problem above, these two sets of limits for the true score will be 58 to 62 and 54.8 to 65.2, respectively.

Exercises

DATA: *Intelligence Test and Social Adjustment Scores for 18 Subjects*

(1) Intelligence: 80 75 71 80 50 64 46 70 64 74 59 84 55 69 86 50 68 65
(2) Adjustment: 146 90 114 77 143 26 88 105 78 44 91 61 44 88 44 182 94 90
(3) Retest Intell.: 82 74 67 80 54 63 44 73 66 77 58 81 50 74 87 53 69 64

Note: These exercises are to be regarded primarily as practice exercises; the results are probably not typical when n is so small.

1. Find the rank-difference correlation between the intelligence test and the social adjustment scores (1 and 2).
2. Repeat Exercise 1 substituting the product-moment method for the rank-difference method. Work out the value of t and interpret in terms of an appropriate null hypothesis and the significance concept.
3. Repeat Exercise 1 using the social adjustment scores and the retest intelligence scores (2 and 3).
4. Using the data of tests 2 and 3, repeat Exercise 2.
5. Using the data of tests 1 and 3, calculate the reliability

coefficient of the intelligence test employed. Make use of the rank-difference method.

6. Repeat Exercise 5, using the product-moment method.

7. Assuming $r = .85$ between the odd and even items on a given test, what is its full-length reliability coefficient?

8. (a) What is the standard error of measurement for a test with an s of 9 and a reliability coefficient of .92? (b) For an obtained score of 70 state, in terms of confidence limits, the probable location of the "true score."

9. Interpret the value of t found in Exercise 2 on the basis of Table 10. Explain any difference in probability values from those based on Table 11.

10. Repeat Exercise 9 using the t value found in Exercise 4.

14
Machine and Chart Correlation Methods; Prediction from Regression Equations; the Phi Coefficient of Correlation

The formula for the product-moment correlation coefficient (r) may be written in many ways, which are simply algebraic transformations of the usual formula (*13.2*) presented in the previous chapter. The most useful variations are based on the formulas which were used in calculating the sum of squares (Σx^2) by the methods of Chapter 4, section 4. One of these applies to machine calculation from original scores and the other is for use with coded grouped scores.

1. Machine Calculation of *r* from Ungrouped Original Scores

If we work with original pairs of scores (X and Y) rather than with deviation scores (x and y), we eliminate the labor of first calculating the means and then the deviations from the means. The drawback to working with original scores, however, is that the numbers involved in the calculation of r sometimes get very large. But most calculating machines do not object. A convenient formula for machine use is

$$r = \frac{n\Sigma XY - \Sigma X \Sigma Y}{\sqrt{[n\Sigma X^2 - (\Sigma X)^2][n\Sigma Y^2 - (\Sigma Y)^2]}} \qquad (14.1)$$

This looks a little rough, but it isn't. With a calculating machine it is possible to get the five Σ's required by the formula *all at the same time*; namely, ΣX, ΣY, ΣX^2, ΣY^2, and ΣXY. (The machine is simply set so that the answer panels do not clear after each pair of scores is entered and squared, but rather record the cumulated sums. ΣX and ΣY then appear in the upper panel and ΣX^2, *2 times* ΣXY, and ΣY^2, appear in the lower panel. We must be careful to correct for this unwanted factor of 2 in the middle sum in the lower panel before entering the Σ's in the formula. The machine is not malicious in introducing this booby trap. It is merely dutifully squaring the sum of X and Y for each pair and correctly recording the familiar $X^2 + 2XY + Y^2$.)

To illustrate the use of this formula, let us apply it to the 30 pairs of psychology grades and intelligence test scores below. From the machine we obtain (if we have made no mistake):

Final Grades (in Percent) in Psychology (X) and Army Alpha Intelligence Scores (Y) for 30 Students

X	Y	X	Y	X	Y
94	194	74	154	83	180
81	165	88	171	84	144
74	118	83	164	80	144
83	150	82	163	86	175
82	139	83	164	82	142
82	144	89	165	92	180
92	185	76	150	75	136
72	139	75	154	77	134
84	186	89	139	84	155
74	156	86	158	62	90

$(n = 30)$

$\Sigma X = 2448$, $\Sigma Y = 4638$, $\Sigma X^2 = 201{,}134$, $\Sigma Y^2 = 730{,}502$, and $\Sigma XY = 381{,}741$. We observe that $n = 30$. Substituting in formula *14.1*, we get

$$r = \frac{30 \times 381{,}741 - 2448 \times 4638}{\sqrt{[30 \times 201{,}134 - (2448)^2][30 \times 730{,}502 - (4638)^2]}}$$

$$= \frac{98{,}406}{\sqrt{[41{,}316][404{,}016]}} = \frac{98{,}406}{10{,}000\sqrt{166.923}} = 0.76$$

This is clearly no job for a subway statistician. But there is a way out, as we shall see in the section which follows.

Before we go on, however, we should observe that the principal work in the machine calculation of r is getting the five Σ's. But these sums serve a double purpose. They can also be used to obtain the two means ($\bar{X}$ and $\bar{Y}$) and the two standard deviations (s_x and s_y), by formulas *3.1*, *4.3*, and *4.4*. These measures are not only useful in themselves, but they are essential for setting up or for interpreting *regression equations*, which are used in making predictions from the correlation (see section 3). In addition to saving the Σ's for later use, it is helpful to keep a record of the three main components in the calculation above; that is,

$$r = \frac{[n\Sigma XY - \Sigma X \Sigma Y]}{\sqrt{[n\Sigma X^2 - (\Sigma X)^2][n\Sigma Y^2 - (\Sigma Y)^2]}}$$

$$= \frac{[A]}{\sqrt{[B][C]}} = \frac{[98{,}406]}{\sqrt{[41{,}316][404{,}016]}} \qquad (14.2)$$

This data makes it easy to calculate *regression coefficients* for the regression equations just mentioned. These will be taken up in section 3.

2. Calculation of *r* from Coded Grouped Data by Means of a Correlation Chart*

As we know from the sections on methods for calculating the mean and the standard deviation in earlier chapters, the best

---

*If the instructor can be sure that his students will always have calculating machines available and in good working condition, he may decide to omit this section, though it is not difficult. The techniques may be applied on buses or on camping trips.

way to avoid large numbers in calculation, or to deal with large numbers of scores, is to group and code our original scores. This is especially true in correlation work, where we may have to deal with large numbers of scores in pairs; for pairs frequently introduce complications. The solution to this problem is the correlation chart. (There are several different forms of correlation chart on the market, but they employ the same principles. One can be made from ordinary graph paper or with a ruler.) The simplest form appears in Table 22, which illustrates the calculation of r for the 30 pairs of psychology grades (X) and Army intelligence scores (Y) that were used for machine calculation above. There are four main steps in the use of the chart:—

Step 1. Filling in the scatter diagram. The scatter diagram is simply a two-dimensional frequency table. The steps are first labeled and the step interval (i) recorded for both the X and the Y variables. (In Table 22, for the sake of clarity, only seven steps have been used. Greater accuracy may be obtained with, say, 11 to 15 steps. An odd number helps to center the 0 reference point but is not required.) *Coding* is accomplished automatically by filling in the marginal X' row and Y' column as indicated, with the 0 point at or near the center. The values of c_x and c_y are recorded. They are, respectively, the midpoints of the steps corresponding to $X' = 0$ and $Y' = 0$. Next *the tallies are entered*. Each tally represents a *pair* of X and Y scores. For example, for the first pair of scores ($X = 94$, $Y = 194$) a tally was put in the upper right hand cell, because 94 falls in the X step range 90–94 and 194 falls in the Y step range 180–194. Two other pairs of scores (92, 185 and 92, 180) fall in these same ranges simultaneously; so two other tallies appear in this same cell. At the other end of the scale we have one tally in the lower left hand cell, representing the score pair (62, 90). The pairs are, of course, usually tallied in the order in which they happen to appear in the data table.

Step 2. Obtaining the sums of the coded scores and the sums of the scores squared. These sums, which are essential for the final calculation, are obtained very easily. This is done by simply filling in columns (1) to (4) and rows (1) to (4) by inspection, in a manner already familiar from Table 7 (p. 43). The sum of the f

Calculation of r from Coded Grouped Data

TABLE 22

Calculation of the Product-Moment Correlation (r) from Coded Grouped Data—A Correlation Chart

Y: Army Alpha Intelligence Test Scores $c_y = 142.0$ $i_y = 15$

X: Final Grades in Psychology (in Percent) $c_x = 77.0$ $i_x = 5$

Y \ X	60–64	65–69	70–74	75–79	80–84	85–89	90–94	(1) f	(2) Y'	(3) fY'	(4) fY'^2	(5) fX'	(6) $Y'fX'$
	X' = -3	-2	-1	0	1	2	3						
180–194 $Y'=3$					//		///	5	3	15	45	11	33
165–179 2			//		/	///		4	2	8	16	7	14
150–164 1			/	//	////	/		10	1	10	10	5	5
135–149 0	/			/	////	/		8	0	0	0	6	0
120–134 -1					/			1	-1	-1	1	0	0
105–119 -2						/		1	-2	-2	4	-1	2
90–104 -3							/	1	-3	-3	9	-3	9
(1) f	1	0	4	4	13	5	3	$30 = n$		27	85	25	63
(2) X'	-3	-2	-1	0	1	2	3			$\sum fY'$	$\sum fY'^2$	$\sum fX'$	$\sum X'Y'$
(3) fX'	-3	0	-4	0	13	10	9	25					
(4) fX'^2	9	0	4	0	13	20	27	73					
(5) fY'	-3	0	0	1	13	7	9	27					
(6) $X'fY'$	9	0	0	0	13	14	27	63					

$\sum$'s

row should of course equal the sum of the f column; for in n X'-Y' pairs there are n X''s and n Y''s and $\Sigma f = n$. (Here $n = 30$.) This provides us with our first accuracy check. The four other important sums which these first four columns and rows produce are $\Sigma fX'$, $\Sigma fX'^2$, $\Sigma fY'$, and $\Sigma fY'^2$. Two of them can be checked for accuracy in the next step.

TABLE 22 (Continued)

Part I. Calculation of r:

$$r = \frac{\Sigma xy}{\sqrt{\Sigma x^2 \cdot \Sigma y^2}} = \frac{A}{\sqrt{B \cdot C}} \tag{14.3}$$

Using the *coded equivalents*, we have

$$\Sigma xy = i_x i_y \left[\Sigma X'Y' - \frac{\Sigma fX' \Sigma fY'}{n} \right]$$

$$= 5 \times 15 \left[63 - \frac{25 \times 27}{30} \right] = 5 \times 15 \times 40.5 \tag{A}$$

And, by formula *4.5*,

$$\Sigma x^2 = i_x^2 \left[\Sigma fX'^2 - \frac{(\Sigma fX')^2}{n} \right]$$

$$= 5^2 \left[73 - \frac{(25)^2}{30} \right] = 5^2 \times 52.2 \tag{B}$$

Also by formula *4.5*

$$\Sigma y^2 = i_y^2 \left[\Sigma fY'^2 - \frac{(\Sigma fY')^2}{n} \right]$$

$$= (15)^2 \left[85 - \frac{(27)^2}{30} \right] = (15)^2 \times 60.7 \tag{C}$$

$$\therefore r = \frac{A}{\sqrt{B \cdot C}} = \frac{5 \times 15 \times 40.5}{5 \times 15 \sqrt{52.2 \times 60.7}} = .72$$

(The $i_x i_y$'s will always cancel.)

TABLE 22 (Continued)

*Part II. Calculation of Means, Regression Coefficients,
and Standard Deviations
(Required for regression equations
and standard error of estimate)*

By formula *3.4*

$$\bar{X} = i_x \frac{\Sigma fX'}{n} + c_x = \frac{5 \times 25}{30} + 77.0 = 81.2$$

$$\bar{Y} = i_y \frac{\Sigma fY'}{n} + c_y = \frac{15 \times 27}{30} + 142.0 = 155.5$$

Regression coefficients

$$\text{For } X \text{ on } Y: b_x = \frac{\Sigma xy}{\Sigma y^2} = \frac{A}{C} = \frac{5 \times 15 \times 40.5}{15 \times 15 \times 60.7} = .222 \quad (14.10)$$

$$\text{For } Y \text{ on } X: b_y = \frac{\Sigma xy}{\Sigma x^2} = \frac{A}{B} = \frac{5 \times 15 \times 40.5}{5 \times 5 \times 52.2} = 2.33 \quad (14.11)$$

By formula *4.6*

$$s_x = \sqrt{\frac{\Sigma x^2}{n-1}} = \sqrt{\frac{B}{29}} = 5\sqrt{\frac{52.2}{29}} = 6.70$$

$$s_y = \sqrt{\frac{\Sigma y^2}{n-1}} = \sqrt{\frac{C}{29}} = 15\sqrt{\frac{60.7}{29}} = 21.7$$

Step 3. Obtaining the sum of the cross-products, $\Sigma X'Y'$. The expression $\Sigma X'Y'$ stands for the sum of the products of the X' and Y' values for every pair of scores. This sum *could* be obtained laboriously by getting *for every tally* the product of its X' and Y' values and then adding all 30 of these products together. In practice we speed up this process by filling in columns (5) and (6) and then getting the sum of the $Y'fX'$ entries in column (6). This equivalent process gives the required sum $\Sigma X'Y'$. *Column (5) is filled in as follows*: In the first row there are 2 tallies in the cell for which $X' = 1$ and 3 tallies in the cell for which $X' = 3$; so the fX' value for this row is $2 \times 1 + 3 \times 3 = 11$. Similarly, the fX' value for the second row is $1 \times 1 + 3 \times 2 = 7$. For the third row $fX' = 2 \times (-1) + 2 \times 0 + 5 \times 1 + 1 \times 2 = 5$. And so it goes. This column is the only one requiring a little concentration. The products of Y' and fX' for each cell in column (6) are easily filled in by inspection. When we have the sum of column (6) we are really ready for the final calculations, but it is smart to check this sum ($\Sigma X'Y'$) by filling in rows (5) and (6) also. The procedure is similar to that used for columns (5) and (6). For example, the fY' entry in the first cell of row (5) is based on the 1 tally in the cell for which Y' is -3; so $fY' = 1 \times (-3) = -3$. And for the third cell of row (5) $fY' = 1 \times (-2) + 1 \times 0 + 2 \times 1 = 0$. And so on. With all six columns and rows filled in we have *accuracy checks* on at least three of our critical sums, four if we count $\Sigma f = n$. We then proceed with reasonable confidence.

Step 4. Making the final calculations. The final calculations are clearly indicated in Table 22, Part I. We first substitute the appropriate Σ's from the chart in the formulas for the *coded equivalents* of Σxy, Σx^2, and Σy^2 (or A, B, and C). (Two of these equivalents are already familiar from Chapter 4.) Then we enter the results in the formula

$$r = \frac{\Sigma xy}{\sqrt{\Sigma x^2 \cdot \Sigma y^2}} = \frac{A}{\sqrt{B \cdot C}} \qquad (14.3)$$

It is interesting to observe that the value of r obtained by the correlation chart (0.72) is not quite the same as the accurate value obtained from the ungrouped original scores (0.76) by formula *14.1*. This is because in grouping scores we let the midpoint of the step represent the whole step. Hence, the larger the step, the greater the potential error. If we had used more steps and a shorter step interval in our chart, the error would have been less. For example, when the same data were used with an 11×11 cell chart (instead of the 7×7 cell chart used in Table 22) r came out 0.74. For larger values of n the correlations by the two methods would also be closer.

3. Predictions from Correlations by Means of Regression Equations and the Standard Error of Estimate

Now that we are equipped to calculate a correlation coefficient, by any of several methods best suited to our data and situation, what can we do with one if we should happen to calculate it? One thing we can do, provided the correlation is fairly high, is to *estimate* individual scores on one of two correlated variables from known scores on the other. (We could, for example, estimate final grades in psychology from intelligence test scores on the basis of our correlation of these variables for a sample of students in the last section.) Such an estimate is made by means of a *regression equation*, which is the equation of a *"regression line."* A regression line does not regress from anything or imply any regression at all.* It is simply the line (assumed to be straight in this discussion†) which best expresses the trend of the

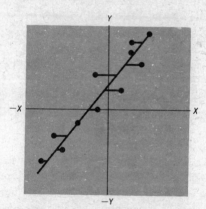

Figure 18. Regression line of X on Y.

---

*Francis Galton, a pioneer in the development of statistical techniques, first used the term in connection with the observed tendency of the offspring of very tall or very short fathers to regress toward the mean height of the population. The term has stuck.

†There are more complicated methods for dealing with curved regression lines which cannot be taken up here.

points on a scatter diagram. For example, in Figure 18 the sloping line is an approximate regression line. Though it does not connect with all the points by any means, it fits the trend pretty well. The short horizontal lines represent the deviations of the X components of the points from this line. The true regression line is not just any line that looks like a good fit. It is "the best fit." This best-fitting line, as usually determined, is the one for which the sum of the squared deviations is as small as possible. It can be shown that the equation of this *"regression line of X on Y"* is, *in deviation scores,*

$$\tilde{x} = b_x y \qquad (14.4)$$

in which $\tilde{x} = (\tilde{X} - \bar{X})$ and $y = (Y - \bar{Y})$. The equation may be written in more convenient terms

$$\tilde{X} = b_x(Y - \bar{Y}) + \bar{X} \qquad (14.5)$$

where $\tilde{X}$ stands for the estimated X score,

Y, $\bar{Y}$, and $\bar{X}$ are familiar

and b_x is "the regression coefficient of X on Y"

Before we discuss this regression coefficient we should note that for a given set of points on a scatter diagram there is usually a second regression line, the regression line of Y on X.* This would be determined in the same manner as the one above, except that in this case the sum of the squared y deviations (*vertical* lines connecting the points with the sloping trend line) would have to be made a minimum. The corresponding *equations for the regression of Y on X* are

$$\tilde{y} = b_y x \qquad (14.6)$$

and
$$\tilde{Y} = b_y(X - \bar{X}) + \bar{Y} \qquad (14.7)$$

Now before we can use these regression equations we must

---

*When $r = 1.00$ or -1.00 the two lines become one, both x and y deviations become 0 (i.e., all points are on the line), and the estimate becomes a certainty!

know how to determine the *regression coefficients* b_x and b_y (which are actually the slopes of their respective regression lines). The most meaningful formulation of these is:—

For X on Y: $$b_x = r\frac{s_x}{s_y} \qquad (14.8)$$

For Y on X: $$b_y = r\frac{s_y}{s_x} \qquad (14.9)$$

These formulas fit in nicely with an r obtained from a correlation chart, or by any other method from which s_x and s_y are readily made available. But *it is simpler to use the equivalent formulas*

$$b_x = \frac{\Sigma xy}{\Sigma y^2} = \frac{A}{C} \quad \text{(Table 22)} \qquad (14.10)$$

$$b_y = \frac{\Sigma xy}{\Sigma x^2} = \frac{A}{B} \quad \text{(Table 22)} \qquad (14.11)$$

Using the results obtained in Table 22 and formula *14.10*, the regression coefficient for X on Y is

$$b_x = \frac{A}{C} = \frac{5 \times 15 \times 40.5}{15 \times 15 \times 60.7} = .222$$

(The i_x and i_y values were left in A and C originally to facilitate cancellation.)

Substituting this value and the values of $\bar{X}$ and $\bar{Y}$ from the same table in the regression equation of X on Y (formula *14.5*), we get

$$\tilde{X} = 0.222(Y - 155.5) + 81.2,$$

which becomes

$$\tilde{X} = 0.222Y + 46.7$$

We can now use this tailor-made equation to *estimate* an individual's final grade in psychology from his intelligence test score. For instance, an Army intelligence score of 180 substituted for Y gives us an estimated grade of

$$\tilde{X} = 0.222 \times 180 + 46.7 = 86.7 \text{ or approximately } 87\%$$

We must be careful not to regard this estimate as a certainty. (Indeed it would be astonishing if all students who got a given score on an intelligence test got the same percentage grade in psychology.) *A regression equation can give us only a best estimate (unless it is based on a perfect correlation).* Since there is a potential error in every estimate, it would be nice to know how large such an error is likely to be, or to know what are the probable limits of this potential error in our estimate.

The standard way of determining these probable limits of error is to calculate what is known as *the standard error of estimate (S)*. This is conveniently given by the formulas

$$S_x = s_x\sqrt{1 - r^2} \qquad (14.12)$$

$$\text{or} \quad S_y = s_y\sqrt{1 - r^2} \qquad (14.13)$$

where S_x is the standard error in the estimate of X by the regression equation of X on Y (formula *14.5*) and S_y is the standard error in the estimate of Y by the regression equation of Y on X (formula *14.7*). If we substitute in the first of these equations the values of s_x and r from Table 22, we get

$$S_x = 6.70\sqrt{1 - (0.72)^2} = 4.6$$

We interpret a standard error of estimate in probability terms similar to those used to interpret the standard error of the mean (see Chapter 7, section 2). That is, in the long run the chances are about 68 in 100 that the actual value of X will lie within one standard error of the estimated value ($\tilde{X}$). Or the chances are about 95 in 100 that the actual value of X lies within 1.96 standard errors of $\tilde{X}$. And $p = .99$ that the actual value will lie within 2.58 standard errors of the estimated value, in the long run.

Now what does this mean in terms of the specific case of our psychology grade estimate (86.7%), obtained by substituting the intelligence score of 180 in our regression equation above? It means that in the long run about 68% of the time the actual psychology grade will be between $(86.7 - 4.6)\%$ and $(86.7 + 4.6)\%$, or between 82% and 91%, approximately. (One of the two actual grades in the original sample, 83, corresponding to $Y = 180$, did fall within this range; the other, 92, fell just outside.) And it means that about 95% of the time the actual grade will be between $(86.7 - 1.96 \times 4.6)\%$ and $(86.7 + 1.96$

× 4.6)%, or between about 78% and 96%. (The actual grades corresponding to $Y = 180$ in the original sample both fell safely within these limits.)

4. Predictions from Machine and Chart Correlations Compared*

It might be interesting to see how much difference, if any, it makes in our estimate of the psychology grade if we use a regression equation based on the more accurate determination of r and the means from the original scores (rather than from coded scores in a simplified scatter diagram), as in section 1. Though we should not expect the difference to be very much, we should, in any case, know *how to put together most simply a regression equation from the Σ's obtained by machine calculation*. We shall use the same formula for *the regression of X on Y*:

$$\tilde{X} = b_x(Y - \bar{Y}) + \bar{X} \qquad (14.5)$$

Using the Σ's already calculated by machine (p. 167), we get

$$\bar{X} = \frac{\Sigma X}{n} = \frac{2448}{30} = 81.6$$

and

$$\bar{Y} = \frac{\Sigma Y}{n} = \frac{4638}{30} = 154.6$$

To obtain the regression coefficient b_x we use the A/C ratio of formula *14.10*, though the values of A, B, and C in the machine formula (*14.2*) are n times as large as those in the chart formula (*14.3*). But since we use *ratios* in calculating both r and the regression coefficients, the n's cancel out and the results are the same. So *in terms of the main components of either the machine or chart formulas for r the regression coefficients are*

For X on Y: $\qquad b_x = \dfrac{A}{C} \qquad (14.10)$

For Y on X: $\qquad b_y = \dfrac{A}{B} \qquad (14.11)$

*This section will be useful, even though section 2 may have been omitted.

Substituting the values of A and C already obtained (p. 167), we get

$$b_x = \frac{98{,}406}{404{,}016} = 0.243$$

We can now write the concrete regression equation for estimating psychology grades (X) from Army intelligence test scores (Y) based on our sample correlation:

$$\tilde{X} = 0.243(Y - 154.6) + 81.6$$

which becomes

$$\tilde{X} = 0.243Y + 44.0$$

For $Y = 180$ we get the estimated psychology grade

$$\tilde{X} = 0.243 \times 180 + 44.0 = 87.7\%$$

This estimated grade is 1% higher than the one obtained by the less accurate method. (Accuracy pays!)

There is one other thing that should interest us in connection with the more accurate determination of r: does the higher correlation (0.76 vs. 0.72) cut down *the standard error of estimate* appreciably? Probably not, unless the correlation is very much higher; but let us check it *to complete the calculation from machine data*. We use the same formula for the standard error of estimate of X on Y:

$$S_x = s_x\sqrt{1 - r^2} \tag{14.12}$$

To get the standard deviation (s_x) we first use the original score formula for Σx^2:

$$\Sigma x^2 = X^2 - \frac{(\Sigma X)^2}{n} \tag{4.4}$$

Substituting the values of ΣX and ΣX^2 already calculated (p. 167) and $n = 30$, we get

$$\Sigma x^2 = 201{,}134 - \frac{(2448)^2}{30} = 1377$$

Then the variance is

$$s_x^2 = \frac{1377}{30 - 1} = 47.5 \tag{4.2}$$

And $s_x = \sqrt{47.5} = 6.90$ (as compared with 6.70 which we got by the rougher methods of the correlation chart)

Substituting this accurate value of s_x and $r = 0.76$ in formula 14.12, we get for the standard error of our estimate of X from the regression equation above

$$S_x = 6.90\sqrt{1 - (0.76)^2} = 4.5$$

This means that about 68% of the time an actual psychology grade corresponding to an intelligence score of 180 would be between $(87.7 - 4.5)$% and $(87.7 + 4.5)$%, or between 83% and 92%, approximately. And about 95% of the time an actual grade would be within 1.96×4.5% of the estimated grade. (Though our two actual grades corresponding to $Y = 180$ fall within the narrower set of limits this time, this is more or less luck.) We have not really sharpened our estimate appreciably by the slight increase in the size of r. The standard error of estimate has merely been reduced from 4.6 to 4.5.

Note: Even if a machine is always available for calculating correlations and regression coefficients (which it isn't), it is useful to know how to make a scatter diagram and to fill in the frequency column and row in the correlation chart. This is so because, if the shapes of the frequency distributions of the two variables, X and Y, are quite different, the possible range of the r values will be considerably less than from 1.00 to -1.00. Though this will not invalidate the prediction procedures we have been discussing, it may cause trouble in more elaborate studies involving an analysis of correlation *patterns* with several variables.

5. Reduction in Error of Estimate as a Function of *r*

The trivial reduction in the error of estimate in the section above for only a slight increase in the size of r was not really surprising. If we look at the formula for the standard error of estimate for a minute, we see that the error of estimate can never be small unless one of two things is true: unless the variability of the original scores (as measured by the standard deviation, s) is small, or unless the correlation is very high. Two extreme cases will make this clear. First, if $r = 0$, then

$$S = s\sqrt{1 - 0^2} = s$$

and the standard error of estimate will be the same as the standard deviation of the original scores, large or small. But if we have a perfect correlation, positive or negative, and $r = 1.00$ or -1.00, then
$$S = s\sqrt{1 - 1^2} = 0$$
In such a case, no matter how large the standard deviation may be, there is no error of estimate and we have predictive certainty. But, unfortunately, since not only Woman, but Man too, is fickle, in practice we never have this happy condition.

In theory, then, the standard error of estimate may range anywhere from 0 (for $r = 1.00$ or -1.00) to the value of the standard deviation (s) of the variable which is being estimated (for $r = 0$). What about the error of estimate for intermediate values of r? How much does an r of 0.50, say, reduce the error below its maximum for $r = 0$? Does it cut it in half? Unfortunately not. It merely reduces the error of estimate by about 13%. The relationship between r and the reduction in the error of estimate is anything but linear. *The reduction in the error of estimate* from its maximum of s may be seen to be equal to $1 - \sqrt{1 - r^2}$. This expression is sometimes called *the index of forecasting efficiency*. The values of this index are recorded in percentage terms in Table 23 for several values of r. This shows us how large r must be in order to produce an important reduction in the standard error of estimate.

We see that in order to reduce the error of estimate by 50% we must have an r equal to nearly 0.87, which is very rare. And for values of r smaller than 0.50 the reduction in the error of estimating individual scores is not very important.

6. The Phi Coefficient of Correlation

In all of our discussions of correlations so far we have been dealing with correlations between two *continuous variables*, X and Y. The score range of a continuous variable may be subdivided into as many categories as we like, or the categories may be made as small as we like so there are no gaps between them. But it often happens that we are interested in seeing if there is a correlation (or association) between two variables each of which can be broken down into just two categories. Such variables are

TABLE 23

Reduction in the Standard Error of Estimate as a Function of r

Correlation r	Standard error of Estimate $s\sqrt{1-r^2}$	Reduction in Error (%) $1-\sqrt{1-r^2}$
1.000	.000s	100.0
.900	.436s	56.4
.866	.500s	50.0
.800	.600s	40.0
.700	.714s	28.6
.600	.800s	20.0
.500	.866s	13.4
.400	.917s	8.3
.300	.954s	4.6
.200	.980s	2.0
.000	1.000s	0.0

called *dichotomous*. There are many examples of dichotomous variables: men and women, right answer and wrong answer, yes and no, Easterner and Westerner, college student and high school student, science major and arts major, above average (or median) and below, Democrat and Republican, Frenchman and German, blond and brunette, natural or artificial, mice and men, etc. It is possible to check an association between two dichotomous variables by means of the phi coefficient of correlation, r_ϕ.

The phi coefficient is a special case of the product-moment correlation, r. In effect we have a 2 × 2 correlation chart instead of one like the 7 × 7 chart used in our illustration in section 2. But it is very easy to fill in a 2 × 2 chart and the r_ϕ formula is equally easy to work with:

$$r_\phi = \frac{ad - bc}{\sqrt{(a+b)(c+d)(a+c)(b+d)}} \quad (14.13)$$

The a, b, c, and d are the frequencies entered in the four cells of the 2 × 2 table, a table like those in the two illustrative cases below. The ad and bc are easy to remember if we are aware of the era we are living in and the one that preceded it. The rather complicated-looking denominator in this formula is really very simple. It does not have to be memorized or filled in in detail. What lies under the radical sign is simply the product of the four marginal totals.

Case I. The answers given to the question "Do you occasionally smoke cigars?" by 100 men and 100 women smokers are classified in the 2 × 2 table below. From this we can find out

Smoke Cigars?

	Yes	No	Σ's
Men Smokers	a 50	b 50	100
Women Smokers	c 10	d 90	100
Σ's	60 +	140 =	200 = n

if any correlation exists between the sex of the smoker and the answers to the question. Substituting the frequencies from the table in the formula, we get

$$r_\phi = \frac{50 \times 90 - 50 \times 10}{\sqrt{100 \times 100 \times 60 \times 140}} = .44$$

Case II. The answers to the question "Do you use lipstick?" given by 100 men and 100 women are classified in the same way.

Use Lipstick?

	Yes	No	Σ's
Men	a 0	b 100	100
Women	c 90	d 10	100
Σ's	90 +	110 =	200 = n

The formula should tell us how close the correlation is between the sex variable and the use of lipstick.

$$r_\phi = \frac{0 \times 10 - 100 \times 90}{\sqrt{100 \times 100 \times 90 \times 110}} = -.91$$

(The minus sign has no special significance; it just depends on how the table was set up.)

If we take the answers to the questions seriously, we conclude that the use of lipstick is much more closely correlated with the sex variable than smoking cigars. And there is nothing much we can do about it.

We should have checked, however, before interpreting the correlations, to set if they were significantly different from 0. (There might have been nothing to interpret.) Let us do it now. In the case of a phi coefficient *a significance test* is made by getting its equivalent in terms of a statistic known as *Chi Square* (χ^2). This is the subject of the next chapter, but we can at least calculate the χ^2 equivalent here by means of the simple formula

$$\chi^2 = nr^2 \qquad (14.14)$$

For our smallest correlation this is

$$\chi^2 = 200(.44)^2 = 38.8$$

If we look at the χ^2 table (p. 188), we find that this is much larger than the tabled value of χ^2 for $d.f. = 1$, which is 6.635 for $p = .01$. This means that we can reject the null hypothesis that the true correlation in the parent population (ρ_ϕ) is 0 with a high degree of confidence. If this is the case for the smaller correlation, it is even more so for the larger one.

Exercises

1. Using the first 10 pairs only of the psychology grades and the intelligence test scores on p. 166, calculate the product-moment correlation by formula *14.1*. (Keep a record of the five Σ's involved for later use. Also keep a record of the brackets A, B, and C, as suggested on p. 167.)

2. Repeat Exercise 1 using the second 10 pairs of scores on p. 166 as your data.

3. Use as your data the first 25 pairs of scores only on p. 166. Tally these in a scatter diagram using the same step labels as those in Table 22. Then: (a) Obtain the sums of the coded scores and the sums of the coded scores squared. (b) Obtain the sum of the cross-products, $\Sigma X'Y'$. (c) Calculate Σxy, Σx^2 and Σy^2. Then calculate r from formula *14.3*.

4. Use all 30 pairs of scores on p. 166 as your data. Construct an 11 × 11 cell scatter diagram, using 77–79 as your central X step and 140–149 as your central Y step. Repeat parts (a), (b), and (c) of Exercise 3.

5. Use the appropriate sums obtained in Exercise 3 to: (a) Find $\bar{X}$ and $\bar{Y}$. (b) Calculate the regression coefficients b_x and b_y. (c) Write the regression equations of X on Y and Y on X.

6. Use the appropriate regression equation from Exercise 5. Then: (a) *Estimate* the intelligence test score of a student who gets a psychology grade of 88% (on the basis of the sample correlated). (b) Find the standard error of this estimate. (c) What are the limits between which we should expect the score to fall about 95% of the time? [Hint: You need standard deviations in (b) and (c).]

7. Use the values of the Σ's and brackets A, B, and C obtained in Exercise 1. Then: (a) Find $\bar{X}$ and $\bar{Y}$. (b) Calculate the regression coefficients b_x and b_y. (c) Write the regression equations of X on Y and Y on X.

8. Repeat Exercise 7 using the values obtained in Exercise 2.

9. Use the appropriate regression equation from Exercise 7 and any other information which you may need from Exercise 1. Then: (a) Estimate the psychology grade of a student whose intelligence test score is 150. (b) Find the standard error of this estimate. (c) What are the smallest limits between which we should expect the grade to fall nearly 99% of the time?

10. On a certain question 50 men answer "Yes" and 50 answer "No." Sixty women answer "Yes" and 40 answer "No." (a) Calculate r_ϕ. (b) Is r_ϕ significantly different from 0?

15
The χ^2 Distribution for Testing Hypotheses

In the determinations of significance in Chapters 7 to 10 the distributions of t and z were made use of in testing the reasonableness of certain null hypotheses. A number of useful kinds of hypotheses may also be tested by means of the χ^2 (chi-square) distribution. Whenever our data constitute a random sample which can be classified into separate categories, we can test the agreement between the observed frequencies and the frequencies to be expected on the basis of some hypothesis by means of the χ^2 test. Several important ways in which this test may be applied are illustrated in this chapter. In the following section the χ^2 statistic is defined and the calculations required for testing "goodness of fit" are illustrated.

1. Calculation of the χ^2 Statistic; Goodness of Fit

The χ^2 technique for testing hypotheses was developed by Karl Pearson in 1900. It is a method of determining whether the differences between the theoretical and the observed frequencies in any number of categories can reasonably be attributed to chance variations in sampling. One of its common uses is testing *goodness of fit* between theory and fact. It involves first the calculation of the statistic χ^2 and then an interpretation of it in terms of probability from a table of the χ^2 distribution.

We may illustrate the procedure conveniently with a concrete problem. Suppose we wish to test the hypothesis that the letter grades in Table 24 are distributed in accordance with the standard normal curve. (If, on the contrary, the observed frequencies deviate significantly from the normal curve, either the class is less brilliant than it should be, or the instructor has exercised a malevolent influence on the grades.) In this table are

TABLE 24

Calculation of χ^2 for Testing Goodness of Fit between Observed and Theoretical (Normal) Distributions of 200 Grades in Elementary Physics

Grades	Frequencies		$(f_o - f_t)$	$(f_o - f_t)^2$	$\dfrac{(f_o - f_t)^2}{f_t}$
	f_o	$f_t{}^a$			
A	9	13.5	-4.5	20.25	1.50
B	38	48	-10.0	100.00	2.08
C	73	77	-4.0	16.00	0.21
D	57	48	9.0	81.00	1.69
E	23	13.5	9.5	90.25	6.69
Σ's	200	200	00.0	...	$\chi^2 = 12.17$

[a]Calculated on the assumption that the "grade base line" $= 1.0\sigma$; values obtained from Table 10 corrected to nearest ½ frequency. For example, since C is in the middle of the distribution, the theoretical range of C's should be from $-0.5z$ to $0.5z$. From Table 10, column (2), we see that 19.2% of the area under the standard normal curve lies between the mean and $0.5z$. So between $-0.5z$ and $0.5z$ we have 38.4%. Therefore, our theoretical frequency (f_t) of C's is 38.4% of the total 200 grades, or 77, approximately. The theoretical A's lie beyond $1.5z$ and the E's below $-1.5z$. The theoretical B's lie between $0.5z$ and $1.5z$. The f_t for the D's is the same as for the B's.

presented the observed frequencies and the theoretical frequencies which would be expected from a normal distribution for $k = 5$ grade categories with "grade base lines" equal to $1s$ (or 1σ). (See Chapter 5, section 3, and Chapter 7.) Once we have these two sets of frequencies, χ^2 can be calculated very easily from the formula

$$\chi^2 = \sum_1^k \frac{(f_o - f_t)^2}{f_t} \qquad (15.1)$$

where f_0 represents the observed frequency and f_t the theoretical frequency, or the frequency which would be expected in terms of the hypothesis to be tested. The Σ_1^k indicates the sum of the $(f_o - f_t)^2/f_t$ terms for all k categories involved in the problem. The steps in the calculation are all clearly indicated in the table. The interpretation of the resulting value of χ^2 is given in the next section.

NOTE: This procedure gives distorted results when the theoretical frequency in any one category is small, say less than 5. When such frequencies occur, it is customary to combine the frequencies of two or more categories to bring them up to 5 at least, or, better, up to 10. It will be observed in Table 24 that the E category, for which the theoretical frequency is even as large as *13.5* has contributed a disproportionate amount to the χ^2 total.

2. Interpretation of χ^2; Use of χ^2 Table

For the interpretation of the χ^2 value found in the problem above we turn to Table 25. This Fisher table of χ^2 is very similar in form to his table of t and, as in the t table, the probability values are expressed in terms of decimals. The column heading *d.f.* stands for *degrees of freedom*, the meaning of which will be explained in the next section. In this particular problem there are four degrees of freedom (see section 3). If we look up our χ^2 value (12.17) in the row for which $d.f. = 4$, we find that it lies between the values for which $p = .02$ and $p = .01$. This means that a value of χ^2 as large as 12.17, or larger, would occur *in the long run* in only 1 or 2 out of every 100 random samples, *if our hypothesis were true*. There are, therefore, fairly safe grounds for rejecting the hypothesis; in this case, the hypothesis that the grades are normally distributed (and some one should speak to

TABLE 25

Fisher's Table of χ^2
(ABRIDGED)

d.f.	p = .99	.98	.95	.90	.10	.05	.02	.01
1	.0001	6.00063	.0039	.0158	2.706	3.841	5.412	6.635
2	.0201	.0404	.103	.211	4.605	5.991	7.824	9.210
3	.115	.185	.352	.584	6.251	7.815	9.837	11.345
4	.297	.429	.711	1.064	7.779	9.488	11.668	13.277
5	.554	.752	1.145	1.610	9.236	11.070	13.388	15.086
6	.872	1.134	1.635	2.204	10.645	12.592	15.033	16.812
7	1.239	1.564	2.167	2.833	12.017	14.067	16.622	18.475
8	1.646	2.032	2.733	3.490	13.362	15.507	18.168	20.090
9	2.088	2.532	3.325	4.168	14.684	16.919	19.679	21.666
10	2.558	3.059	3.940	4.865	15.987	18.307	21.161	23.209
11	3.053	3.609	4.575	5.578	17.275	19.675	22.618	24.725
12	3.571	4.178	5.226	6.304	18.549	21.026	24.054	26.217
13	4.107	4.765	5.892	7.042	19.812	22.362	25.472	27.688
14	4.660	5.368	6.571	7.790	21.064	23.685	26.873	29.141
15	5.229	5.985	7.261	8.547	22.307	24.996	28.259	30.578
16	5.812	6.614	7.962	9.312	23.542	26.296	29.633	32.000
17	6.408	7.255	8.672	10.085	24.769	27.587	30.995	33.409
18	7.015	7.906	9.390	10.865	25.989	28.869	32.346	34.805
19	7.633	8.567	10.117	11.651	27.204	30.144	33.687	36.191
20	8.260	9.237	10.851	12.443	28.412	31.410	35.020	37.566
21	8.897	9.915	11.591	13.240	29.615	32.671	36.343	38.932
22	9.542	10.600	12.338	14.041	30.813	33.924	37.659	40.289
23	10.196	11.293	13.091	14.848	32.007	35.172	38.968	41.638
24	10.856	11.992	13.848	15.659	33.196	36.415	40.270	42.980
25	11.524	12.697	14.611	16.473	34.382	37.652	41.566	44.314
26	12.198	13.409	15.379	17.292	35.563	38.885	42.856	45.642
27	12.879	14.125	16.151	18.114	36.741	40.113	44.140	46.963
28	13.565	14.847	16.928	18.939	37.916	41.337	45.419	48.278
29	14.256	15.574	17.708	19.768	39.087	42.557	46.693	49.588
30	14.953	16.306	18.493	20.599	40.256	43.773	47.962	50.892

For larger values of d.f., the expression $\sqrt{2\chi^2} - \sqrt{2d.f. - 1}$ may be used as a normal deviate with unit standard error.

This table was taken with the consent of the author and publisher from R. A. Fisher's *Statistical Methods for Research Workers*, published by Oliver and Boyd, Edinburgh.

the physics instructor). Repeated rejections under similar circumstances would result in false rejections only about once or twice in every 100 times. *In general, if p lies between .10 and .90 there is no reason to reject the hypothesis being tested; but if p is less than .02* (as in our problem) the hypothesis is pretty dubious. As in the case of interpreting t, the criteria for the acceptance or rejection of hypotheses are arbitrary, but a conventional rule draws the line at $p = .05$ and regards a hypothesis as probably false for values of p smaller than this (that is, for values of χ^2 larger than the value corresponding to a p of .05). It should be pointed out that excessively small values of χ^2, which result in large values of p (say, larger than .95), occur just as rarely on the basis of chance variations as very large values of χ^2, for *true* hypotheses. Therefore, such very small values of χ^2 also constitute grounds for the rejection of hypotheses, though this fact is sometimes overlooked.

(Though the χ^2 table does not give values beyond $d.f. = 30$, reasonably accurate values of p may be found for larger values of $d.f.$ by working out the value of the expression $\sqrt{2\chi^2} - \sqrt{2d.f. - 1}$ and regarding it as normally distributed about zero, with a standard error equal to 1; that is, we may find approximate values of p by looking up the values of this expression in the z distribution, Table 10. If this quantity exceeds 1.96 for large values of $d.f.$, or 1.65 for values not much larger than 30, we get a value of p small enough to cast doubt on the hypothesis.)

It should be observed that, though the χ^2 test has many special uses, it is not as rigorous as tests based on the t distribution, especially when applied to discontinuous distributions, as it usually is. Its weakness in dealing with small theoretical frequencies has already been mentioned. Another defect is its failure to take into account the *direction* of the divergence of observed frequencies from the expected values; for the minus signs in the $(f_o - f_t)$ terms disappear on squaring. A method of correcting for discontinuity is given in section 5.

3. Degrees of Freedom in χ^2 Tests

In Table 24 (section 1) the χ^2 statistic was calculated for five categories (grades A–E). In section 2 it was stated without explanation that for this problem there were four degrees of freedom. What does this mean? *In general, the number of degrees of freedom (d.f.) is the number of theoretical frequencies that can be assigned arbitrarily;* that is, the number of cells in a table for calculating χ^2 that can be filled in arbitrarily. In Table 24 there is one condition imposed on the f_t column; namely, that it shall have the *same total* as the f_o column (200). We could arbitrarily fill in the frequencies in four of the cells of this column with an endless variety of frequency patterns, conforming to an equal number of hypotheses, but we would have no freedom of choice about the entry in the fifth cell; for it would be definitely determined by the requirement that the total equal 200. In cases of this type, *where we are dealing with a single column or row of frequencies for k categories, and when the only condition is a fixed total frequency, d.f. = k − 1.*

(Some authorities would argue that in the problem of Table 24 two other conditions, or restrictions, are implicitly imposed on the distribution in the f_t column; namely, that the mean shall be in the middle of the C category, say at 75%, and that the s for the f_t column shall equal the s for the f_o column. It would then be in order to apply the rule that *the number of degrees of freedom is reduced by one for each constant derived from the observed frequencies;* in this case, $d.f. = k − 3$. However, the practice followed above is the more common.)

Table 24 may be classified as a *k × 1 table,* implying that the theoretical frequencies fall into a *single row, or column,* of k cells. Let us next consider the determination of *d.f.* for some common forms of *multiple row tables.* The simplest case would be a 2 × 2 *table,* often called a fourfold table. If the only restriction imposed on such a table of theoretical frequencies were that the total for all four cells should be fixed, the *d.f.* value would be $k − 1$ or 3, as in the case of a $k × 1$ table with a similar single restriction; but more commonly we have the following requirement: that the totals for *both* rows *and* columns be fixed. Figure 19 is an example of this. A moment's consideration will show that for such a table $d.f. = 1$; for a frequency entered in any one of the four cells at once determines all the others. For example, the 20 entered in the upper left-hand cell compels us to enter the 80 in the upper right-hand cell (100 − 20 = 80); and this same 20 also determines the 15 in the lower left-hand cell (35 − 20 = 15); consequently, 35 is the only possible entry in the lower righthand cell (50 − 15 = 35).

When we have to deal with larger frequency tables we can work out the value of $d.f.$ on similar principles. There is a helpful formula, which applies when such tables are restricted as in Figure 19; that is, with the totals for both rows and columns fixed, but with no other restrictions: for such cases $d.f. = (r − 1) \times (c − 1)$, where r stands for the number of rows and c for the number of columns. For example, if we had a 4 × 5 table with its 4 rows and 5 columns so restricted, $d.f.$ would be 3 × 4 = 12. This formula applied to the 2 × 2 table of Figure 19 will give us the correct value for $d.f.$ also. A problem involving this type of table is discussed in the next section.

		Σ's
20	80	100
15	35	50

Σ's   35   115   150

Figure 19.

4. The Use of χ^2 in Tests of Independence (and Association)

In the first two sections we saw how the χ^2 statistic could be used to test goodness of fit. It has another important application in *tests of independence*. In this class of tests we test the hypothesis that two variables or traits are independent of each other. Let us first illustrate the procedure with a simple 2 × 2 table, which is used whenever the frequencies for each variable can be fitted into just two categories.

Table 26 gives the observed frequencies of right and wrong answers on a certain intelligence test item for a random sample of science and arts majors. We wish to test the hypothesis that performance on this item is independent of the field of academic

TABLE 26

Calculation of χ^2 for a 2 × 2 Table in a Test of Independence

(TESTING HYPOTHESIS THAT RESPONSE ON AN INTELLIGENCE TEST ITEM IS INDEPENDENT OF FIELD OF TRAINING)

	Right	Wrong	Σ's
Science Majors	a 9 (6)	b 6 (9)	15
Arts Majors	c 15 (18)	d 30 (27)	45
Σ's	24 (+)	36 (=)	60 (n)

Method I

A. *Determination of Theoretical Frequencies*

	Cell	Calculation		f_t
	a	$\dfrac{15 \times 24}{60}$	=	6
	b	15 − 6	=	9
	c	24 − 6	=	18
	d	45 − 18	=	27
or	d	$\dfrac{45 \times 36}{60}$	=	27 (check)

B. *Calculation of χ^2*

Cell	$(f_o - f_t)$	$(f_o - f_t)^2$	$(f_o - f_t)^2/f_t$
a	3	9	1.50
b	−3	9	1.00
c	−3	9	0.50
d	3	9	0.33
Total	0.00		$\chi^2 = 3.33$

TABLE 26 (continued)

Method II

Calculation of χ^2 Directly from Observed Frequencies by Formula:

$$\chi^2 = \frac{n(ad - bc)^2}{(a + b)(c + d)(a + c)(b + d)} \qquad (15.2)$$

where a, b, c, and d are the observed frequencies, n is the grand total (i.e., their sum), and the denominator is simply the product of the four marginal totals. In this probelm,

$$\chi^2 = \frac{60(9 \times 30 - 6 \times 15)^2}{15 \times 45 \times 24 \times 36} = 3.33 \text{ (same result)}$$

For Both Methods

$\chi^2 = 3.33 \quad d.f. = 1 \quad p$ lies between .05 and .10

---

training. There are two methods for calculating χ^2. In *Method I*, the more general one, we first determine the theoretical frequencies from the marginal totals. We can readily observe that 15/60 of the entire group were science majors, that 45/60 were arts majors, and that there were 24 right answers altogether. Therefore, if correctness of response is uninfluenced by the field of training, the chance distribution of right answers would be the same as the proportion of subjects in each subgroup. We should then expect 15/60 × 24, or 6, of the science majors and 45/60 × 24, or 18, of the arts majors to get right answers. The expected frequencies for the wrong answers may be determined by subtraction from the marginal totals. These theoretical frequencies are entered in the parentheses in each cell. In practice, *it is necessary to calculate only one theoretical frequency by using the method of proportion*, as above; the other three are determined by subtraction from the marginal totals, as in the table. (Hence, $d.f. = 1$, as in Figure 19 in the preceding section.) However, f_t *can always be calculated for any cell in such a table (a "contingency table"), no matter how large, by dividing the products of the two corresponding marginal totals (the horizontal and the vertical) by the grand total* (n), as was done in the table for cells a and d.

After the theoretical frequencies have been determined, we must then go through the usual steps in calculating χ^2 (part B under Method I in table).

Method II (Table 26) is a great timesaver; for it gives us χ^2 directly in one formula, making it *unnecessary to determine the theoretical frequencies*. Though the two methods give us identical results, Method I is not forced into the trash pile by the streamlined procedure of Method II; for it is a method which also applies to larger tables, and it is sometimes desirable to know in just what cells the largest discrepancies between observed and theoretical frequencies occur. *In interpreting the value of p obtained in this illustration by either method (p lies between .05 and .10), we would retain the hypothesis of independence with a fair degree of confidence and conclude that the responses to the test item under consideration do not differ significantly from those we should expect if they were independent of the fields of training represented by our samples.* There was, in other words,

TABLE 27

The χ^2 Test of Independence Applied to the Hypothesis that High School Grades Are Independent of Social Adjustment
(3×5 Table)

Social Adjustment	Average Grade					Σ's
	A	B	C	D	E, F	
Good	1A 15 (10.9)	1B 45 (32.7)	1C 60 (65.5)	1D 25 (30.0)	1E 5 (10.9)	150
Satisfactory	2A 20 (21.8)	2B 65 (65.5)	2C 140 (130.9)	2D 60 (60.0)	2E 15 (21.8)	300
Poor	3A 5 (7.3)	3B 10 (21.8)	3C 40 (43.6)	3D 25 (20.0)	3E 20 (7.3)	100
Σ's	40	120	240	110	40	550 (n)

A. Determination of Theoretical Frequencies

Cell	Calculation	f_t	Cell	Calculation	f_t
1A	$\dfrac{150 \times 40}{550} =$	10.9	2A	$\dfrac{300}{550} \times 40 =$	21.8
1B	$\dfrac{150 \times 120}{550} =$	32.7	2B	$\dfrac{30}{55} \times 120 =$	65.5
1C	$\dfrac{150 \times 240}{550} =$	65.5	2C	$\dfrac{6}{11} \times 240 =$	130.9
1D	$\dfrac{150 \times 110}{550} =$	30.0	2D	$0.545 \times 110 =$	60.0
1E	$\dfrac{150 \times 40}{550} =$	10.9	2E	$0.545 \times 40 =$	21.8
	(check)	150.0		(check)	300.0

3A $40 - (10.9 + 21.8) = 7.3$ 3D $110 - (30.0 + 60.0) = 20.0$
3B $120 - (32.7 + 65.5) = 21.8$ 3E $40 - (10.9 + 21.8) = 7.3$
3C $240 - (65.5 + 130.9) = 43.6$ (check) 100.0

B. Calculation of χ^2

$\chi^2 = (15 - 10.9)^2/10.9 + (45 - 32.7)^2/32.7 \ldots + (5 - 10.9)^2/10.9$
$\qquad + (20 - 21.8)^2/21.8 \ldots + (15 - 21.8)^2/21.8$
$\qquad + (5 - 7.3)^2/7.3 \ldots + (20 - 7.3)^2/7.3 = 44.30$

d.f. $= (r - 1)(c - 1) = (3 - 1)(5 - 1) = 8$
p is much less than .01

---

probable independence of the field of training, but some conformity with the theoretical expectation based on chance.

In Table 27 we have another illustration of a χ^2 test of independence, this time for *a 3 × 5 table*. Here we are testing the hypothesis that high school grades are independent of social adjustment. The calculations for determining the theoretical frequencies are given in full, and methods of checking the results are suggested. The f_t values for the first two rows are calculated by *the method of proportion* explained under Method I (page 193) for a 2 × 2 table. This is obvious for the first row, but it may take a moment's consideration to see that it is also true for the second row; for here the pattern has been changed slightly to suggest a method of speeding up the calculations. (Obviously a slide rule or a calculating machine would be helpful in all these calculations.) The last cell in each of these rows could have been determined *by subtraction* from the marginal totals, if more convenient. The frequencies for the entire last row were determined in this way. This procedure may save time; it was used here to emphasize the fact that these bottom and end cells may *not* be filled in arbitrarily and, hence, the $d.f.$ formula which applies here is $d.f. = (r - 1)(c - 1) = (3 - 1)(5 - 1) = 8$. Once the f_t values are determined, the χ^2 calculation is carried out in the usual way, as suggested in outline form. If we look up the resulting value of χ^2 (44.30) in Table 25 for $d.f. = 8$, we find that p is much less than .01. This means that we can reject our hypothesis with great confidence; in other words, if our sample is typical, high school grades are *not independent* of social adjustment, but *there is some association* between the two variables.

One of the weaknesses of the χ^2 technique is that it does not tell us the *degree* of such association. However, we can get some light on the relationship if we examine a table which gives the values of $(f_o - f_t)^2/f_t$ for each cell of the original frequency table. Table 28 gives us these values for the data of Table 27. It is clear from the row totals that the heaviest contributions to χ^2 were made by the groups which showed good and poor adjustment. And from the column totals we see that the deviations

TABLE 28

Values of $(f_o - f_t)^2/f_t$ Calculated from the Data of Table 27, Suggesting Why Grades Are Not Independent of Social Adjustment

Social Adjustment	Average Grade					Totals
	A	B	C	D	E, F	
Good	1.54	4.63	0.46	0.83	3.19	10.65
Satisfactory	0.15	0.00	0.63	0.00	2.12	2.90
Poor	0.72	6.39	0.30	1.25	22.09	30.75
Totals	2.41	11.02	1.39	2.08	27.40	$\chi^2 = 44.30$

from the hypothesis of independence are most marked in the B and E, F grades. The association, or relationship, between adjustment and grades stands out most markedly, then, in four cells: the B cells for both the good and the poor adjustment groups, and the E, F cells for both the good and the poor adjustment groups. As has been pointed out before, we cannot determine the *direction* of the deviation of observation from hypothesis from χ^2 data; but if we re-examine Table 27 with our attention focused on these four cells, it is easy to see that the hypothesis of independence was exploded primarily because those who showed good adjustment got far more than their "quota" of B's and less than their quota of E's and F's, and those who showed poor adjustment got far less than their quota of B's and far more than their quota of E's and F's.*

5. Correction for Discontinuity

A slight modification of the χ^2 formula has been suggested by Yates to correct for the fact that the χ^2 distribution is continuous, whereas the observed frequencies are discrete, or discontinuous. This refinement is recommended *only for applications in which there is just one degree of freedom*. This *usually* means when there are only two categories, when $k = 2$. But it would also apply to a 2 × 2 table like Table 26 where both row and column totals are fixed; and where, therefore, $d.f. = 1$.

Formula *15.1*, with the correction for discontinuity, becomes

$$\chi_c^2 = \sum_1^2 \frac{(|f_o - f_t| - 0.5)^2}{f_t} \qquad (15.3)$$

where the subscript c indicates that χ^2 is "corrected." The only

---

*As this discussion suggests, χ^2 tests of independence may sometimes be regarded as *tests of association*. Mathematically, the two are identical. The principal difference between them lies in the phrasing of the hypothesis and in the interpretation of χ^2. For example, if we set up a hypothesis that two variables are independent, we reject this hypothesis when we get small values of p and conclude that there is some association between them, as in the problem of Tables 27 and 28, just discussed. But if we set up a hypothesis that some association exists between two variables, we retain this hypothesis for moderately large values of p.

changes from formula *15.1* are that we must use the *absolute* values of $(f_o - f_t)$ and that 0.5 is subtracted from this before squaring. The 2 over the Σ implies that this correction formula usually involves *just two terms*, one for each of two categories; so that $k - 1 = 1$ d.f. But if we apply it to the four cells of Table 26 (because d.f. = 1), there will be *four terms*. The values of these terms are somewhat smaller than those obtained by formula *15.1* in *Method I*. They are 1.04, 0.70, 0.35, and 0.23; and χ_c^2 becomes 2.32.

A similar correction for discontinuity may be applied to the illustration used in *Method II* in Table 26. Formula *15.2* corrected for discontinuity becomes

$$\chi_c^2 = \frac{n(|ad - bc| - n/2)^2}{(a+b)(c+d)(a+c)(b+d)} \qquad (15.4)$$

Using the observed frequencies and marginal totals from this same table, we get

$$\chi_c^2 = \frac{60(9 \times 30 - 6 \times 15 - 60/2)^2}{15 \times 45 \times 24 \times 36} = 2.31$$

Using d.f. = 1 in Table 25, we find that $2.31 < 2.71$, the tabled value for $p = .10$. This enables us to retain the hypothesis that responses to the test item are independent of the fields of training, with an even greater degree of confidence than we did on the basis of the uncorrected χ^2 value. Though the use of the corrected χ^2 made little difference in this case, there are borderline cases where the corrected value might well change our decision.

Exercises

1. Replace the values in the f_o column of Table 24 with the values 11, 40, 87, 52, and 10; then calculate χ^2. With the aid of Table 25 and an appropriate hypothesis, interpret the result.

2. Using the value of χ^2 obtained in Table 27 interpret *on the assumption* that d.f. = 32. (This requires the use of the

special formula below Table 25; then look up p in Table 10.)

3. (a) If the illustrative problem of Table 24 had involved six categories, how many degrees of freedom would there have been? (b) In a table of three rows of six columns each what will be the value of $d.f.$ when: (1) only the grand total is fixed; (2) only the row totals are fixed; (3) both row and column totals are fixed?

4. The mean height of 200 boys is 60.0 inches; the standard deviation is 4.0 inches. The observed distribution of heights is as follows:

Height category	f_o
Above 66.0	15
64.0–66.0	15
60.0–64.0	70
56.0–60.0	62
54.0–56.0	20
Below 54.0	18

(a) Express the height categories in terms of z scores. (b) With the aid of Table 10 calculate the theoretical frequencies for each of these categories on the assumption that height is normally distributed. (c) With the aid of the χ^2 statistic, test the hypothesis that the heights of these boys are normally distributed.

5. In a group of 100 boys, 30 are classified as "athletes" and 70 as "nonathletes." Of the 30 athletes, 20 show "good adjustment" and 10 show "fair adjustment." Of the nonathletes, 30 show good adjustment and 40 show fair adjustment. (a) Set up this information in a 2 × 2 table with the proper headings and marginal totals. (b) Calculate the theoretical frequencies on the assumption that "adjustment" is independent of athletic ability. (c) Calculate χ^2 by Method I of Table 26. (d) Calculate χ^2 by Method II. (e) Interpret in terms of the hypothesis implied in part (b).

6. Using the correction for discontinuity, repeat Exercise 5 (d). Does the result change the interpretation in 5 (e)? Explain.

The χ^2 Distribution for Testing Hypotheses

7. Assume the following changes in the *observed* frequencies in the top row of Table 27 and consequently in the totals:

	A	B	C	D	E, F	Σ's
Good:	10	30	35	20	5	100
Σ's:	35	105	215	105	40	500 (n)

(a) Calculate the theoretical frequencies on the basis of the hypothesis of the independence of grades and social adjustment.
(b) Calculate χ^2. (c) Interpret.

8. Make a table of the values of $(f_o - f_t)^2/f_t$ for the data of Exercise 7 and pick out the cells that contributed most to χ^2. Interpret these findings in terms of the independence or the association of the two variables of this problem.

16
Significance Tests Based on the Order of Scores

The classical t and F significance tests discussed in earlier chapters both involved assumptions about the normal form and other characteristics, or parameters, of the larger populations from which random experimental samples were drawn. These tests are commonly called *parametric tests*. The calculations take into account the exact size of the scores involved in the various groups compared. There are a number of somewhat simpler significance tests, based on simpler probability theory, which do not depend on the exact size of individual scores, but rather upon their general position or the order in which they appear in a scale. Since they usually involve fewer assumptions about the characteristics of the parent populations, they are sometimes called *nonparametric tests*. But since their essential characteristic is the use of the order, rather than the exact size, of the measures involved, it is more appropriate to describe them as *order tests*.

In this final chapter we shall deal with two such *order significance tests*. The first is based on the rank-order of the scores in two or more groups to be compared. The second is simply based on the position of the scores in the groups to be compared above or below the grand median. Both of these tests may be evaluated by means of the χ^2 distribution of the last chapter.

1. The Kruskal-Wallis Rank Test of Significance

When our data consist of ranks the t and F significance tests usually should not be used. But Kruskal and Wallis (among others) have developed a significance test intended for use with the rank-order of scores. It can be used with two or more experimental groups. When used with more than two groups it is *in effect a kind of simple analysis of variance*. If the method is used with only two groups, it is *roughly the equivalent of a t test (or z test)* for the difference between two means.

The test has in some circumstances certain advantages over the t and F tests, but it also has certain disadvantages. Among *its advantages* are: (1) It is appropriate for use when the rank-order of the scores actually has more psychological meaning than their exact size (as, for example, when subjects or objects have been ranked by judges on some quality). (2) It is relatively insensitive to differences in variance; whereas the classical t and F tests assume homogeneity of variance. (3) The calculations are relatively easy when N is not large, and the calculations are nearly as easy for groups of unequal size as for groups of the same size. However, the Kruskal-Wallis Rank Test is *no general substitute for the classical parametric significance tests* because: (1) In using this test when the t and F tests are applicable there may be some loss of power but not a great deal. (2) There is no simple equivalent of the Scheffé test for locating the experimental effects. (3) The test is not designed for two-factor analysis. (4) Unless rank-order has more psychological meaning than the exact size of the scores, some information is sacrificed in using ranks. (5) When there are many tied ranks, the test loses accuracy unless correction procedures are used.

The Kruskal-Wallis significance test is based on a statistic called H, which has a distribution very similar to χ^2 for $d.f. = k - 1$, where k is the number of experimental groups. The χ^2 table can, therefore, be used to evaluate H. The formula for H is

$$H = \frac{12}{N(N+1)} \sum_1^k \frac{T_k^2}{n_k} - 3(N+1) \qquad (16.1)$$

where N is the total number of scores in all k groups, n_k is the number of scores in the kth group, and T_k is the total of *the ranks* in the kth group. Of course, k takes values $1, 2, \ldots k$. Scores are ranked for all groups considered as one large group. The use of the formula can best be illustrated with concrete data. It may be interesting to use the same data used earlier in our illustration of simple analysis of variance, p. 122, and thus to see how the results from the two methods compare. The scores from the earlier illustration are repeated in Table 29, along with their ranks. In this case the number of scores in each of the k groups is the same ($n_1 = n_2 = n_3 = 5$); but *this is not necessary*, as is made clear in Table 30, with the same data regrouped.

The *four steps* involved in the use of the method are clearly indicated in the two tables, which are self-explanatory. However, *Step 1* may be supplemented by the use of a very simple *accuracy check on ranking and summing the ranks*. This is illustrated for Table 29 as follows: The total of all ranks is given by

$$T = \frac{N(N+1)}{2} = \frac{15(16)}{2} = 120$$

From our group totals we obtained

$$T = \sum_1^k T_k = 40 + 20 + 60 = 120$$

No mistake so far! For Table 30, though the groups are rearranged, N remains the same; and, therefore, T is the same. Since the sum of the group totals adds up to 120 again, we are still with it.

TABLE 29

The Calculation of H from Ranked Scores for Three Groups
(DATA FROM TABLE 14)

	Punishment (1)		Reward (2)		Control (3)	
	Score	Rank	Score	Rank	Score	Rank
	3	10.5	5	5.5	1	14
	5	5.5	7	2	3	10.5
	4	8	6	3.5	2	12.5
	6	3.5	8	1	4	8
	2	12.5	4	8	0	15
Group Totals, T_k	40.0		20.0		60.0	
T_k^2/n_k	320.0		80.0		720.0	

Step 1. Scores are ranked for all groups considered as one large group, with $N = n_1 + n_2 \ldots + n_k$. Rank 1 goes to the largest score, N to the smallest. Tied scores are given average of ranks. *Ranks are summed.* (See text for accuracy check.)

Step 2. Calculation of

$$\sum_1^k \frac{T_k^2}{n_k} = \frac{(40)^2}{5} + \frac{(20)^2}{5} + \frac{(60)^2}{5}$$

$$= 320.0 + 80.0 + 720.0 = 1120.0$$

Step 3. Substitute values of N (15) and $\Sigma_1^k \dfrac{T_k^2}{n_k}$ *in the formula:*

$$H = \frac{12}{N(N+1)} \sum_1^k \frac{T_k^2}{n_k} - 3(N+1) \qquad (16.1)$$

$$= \frac{12(1120)}{15(16)} - 3(16) = 8.0$$

Step 4. Interpret H from χ^2 table (p. 188). H_o is that the k samples come from the same, or identical, populations. Using $d.f. = k - 1 = 2$, we find that $\chi^2 = 9.21$ for $p = .01$ and 7.82 for $p = .02$. Since our obtained H was $8.0 > 7.82$, we reject H_o at the $p < .02$ level.

We can say a word more about the *interpretation in Step 4*. The null hypothesis, H_o, is that all k groups are random samples from the same population, or identical populations. For both

TABLE 30

*The Calculation of H from Ranked Scores
for Two Groups of Unequal Size*
(DATA FROM TABLE 14 REGROUPED)

Incentive Groups Combined (1–2)		Control Group (3)	
Score	Rank	Score	Rank
3	10.5	1	14
5	5.5	3	10.5
4	8	2	12.5
6	3.5	4	8
2	12.5	0	15
5	5.5		
7	2		
6	3.5		
8	1		
4	8		
Group Totals, T_k	60.0		60.0
T_k^2/n_k	360.0		720.0

Step 1. Scores are ranked as in Table 29 *and ranks are summed.* (See text for accuracy check.)

Step 2. Calculation of

$$\sum_1^k \frac{T_k^2}{n_k} = \frac{(60)^2}{10} + \frac{(60)^2}{5} = 360 + 720 = 1080$$

Step 3. Substitute values of N (15) and $\sum_1^k \frac{T_k^2}{n_k}$ *in formula 16.1:*

$$H = \frac{12(1080)}{15(16)} - 3(16) = 54 - 48 = 6.0$$

Step 4. Interpret H from χ^2 *table* (p. 188). Using $d.f. = k - 1 = 1$, we find that $\chi^2 = 6.635$ for $p = .01$ and 5.41 for $p = .02$. Since our obtained H was $6.0 > 5.41$, we reject H_o at the $p < .02$ level and conclude that the samples come from different populations, with some confidence.

---

tables H was large enough so that the null hypothesis was rejected at the $p < .02$ level; that is, with a fairly high degree of confidence. The conclusion in both cases is that the populations are not identical. We are primarily concerned with differences between the means (or the medians) of the populations. Though the H formula involves no measures of central tendency *directly*, it was developed in such a way that we may safely conclude that *significant χ^2 values are primarily the result of significant differences between means (or medians)*.

Where only two groups are involved, as in Table 30, we can get a good clue as to *the direction of the difference between population means*. When H equals or exceeds the tabled value of χ^2 for $p = .05$ (for $k - 1 = 1$ $d.f.$), as in this illustration, it is a fairly safe bet that *in the long run the larger population mean (or median) corresponds to the sample with the smaller average rank*. In this case the *average* rank (not the sum) for the combined "incentive" groups was $60/10 = 6.0$, and for the control group it was $60/5 = 12.0$. This means that the incentives had a significant positive reinforcing effect.

2. The H Test Compared with the F and t Significance Tests

Since we used the same experimental data in our H test illustrations that we used earlier in illustrating simple analysis of variance (Table 14), we can make two specific comparisons. First in using the H test as a kind of simple analysis of variance, that is *as a preliminary check on a possible over-all experimental effect somewhere in our k groups*, we came to the same conclusion that we did in using the more elaborate F test. The only difference was that with the F test we were able to reject the null hypothesis with a little more confidence: $p < .01$ as compared with $p < .02$ for the H test (Table 29). Under the circumstances this made no difference in our decision; but if our obtained values for both F and H had been a little smaller, we might have made two different decisions. We might still have rejected H_o for F, but not for H. In other words, the F test was a little more sensitive. In a borderline case it has a little more power to protect us against a Type II error.

The second comparison involved *the use of the H test with just two groups,* groups of unequal size (Table 30), as *a rank equivalent of a t test.* We got a value of H which enabled us to reject the null hypothesis at the $p < .02$ level. In our original illustration (Chapter 11) we did not have the equivalent t test result; but we can easily work it out now, using Fisher's formula for small samples of unequal size, formulas *8.1* and *8.4*, we get $t = 3.12$. From Table 11, for $d.f. = 10 + 5 - 2 = 13$, we find that $t = 3.01$ for $p = .01$. Since $3.12 > 3.01$, we reject H_o at the $p < .01$ level. So again we reach the same conclusion with the H rank test and the exact score t test. But again the t test is a little more sensitive and gives us a little more power.

In general, however, *for limited uses, the H test has been found to compare rather favorably with the F and t tests.* Its uses are limited because there is no efficient equivalent of the Scheffé follow-up test and no simple application to two-factor analysis. But if the preliminary H test indicates that there is an over-all experimental effect somewhere in our k groups (as in Table 29), we could, of course, inspect the *average* rank (not the rank sum) for each of our k groups and then *apply the H test all over again to any two groups that show big differences in average ranks. If we do this, we must re-rank the scores each time using a new value of N.* (The new N must be the number of scores in *just the two groups* being compared.) For example, if we wanted to compare Group 2 with Group 3 in Table 29, N would become 10. Score 8 would still be ranked 1, but score 0 would have its rank changed to 10. Then all the calculations would have to be done over again.

So, though we may save some calculation time on the preliminary analysis by the use of the H test, we lose out on the follow-up procedures for locating specific differences between groups and on all the special uses of two-factor analysis. And we sometimes lose a little power.

3. The Median Test

The median test is one of the simplest of the significance tests based on the order of the scores in experimental data, rather than on their exact size. It makes use of simple directional deviations from the median (positive or negative), rather than the exact size of the deviations from the mean which are used in the t and F tests. It does not require the assumption that the populations from which random experimental samples are drawn are normally distributed. It is, therefore, *sometimes used when the assumption of normality does not seem justified*. *The calculations involve four steps* which are all fairly easy. We shall illustrate these four steps with 20 experimental group scores and an equal number from a control group, $n_1 = n_2 = 20$.

20 Experimental and 20 Control Group Scores

E: 12 11 11 10 9 9 8 8 7 7 7 7 6 6 | 5 5 5 3 3 1
C: 10 8 6 6 6 6 | 5 5 5 5 5 4 4 4 3 3 2 2 1 0

Step 1: *Find the grand median for all scores in both groups combined into one group.* ($n = n_1 + n_2 = 20 + 20 = 40$.) For our data combined from both groups we get the following frequency table, from which the grand median is easily found to be 5.5.

Frequency Table for Combined Data
   ($n = 40$)

X	f
12	1
11	2
10	2
9	2
8	3
7	4
6	6
5	8
4	3
3	4
2	2
1	2
0	1

$n/2$ or 20 cases lie above and below the upper limit of score 5; so the *grand median is 5.5*.

Step 2: *For each group count the number of scores above and below the grand median and enter the count in a 2 × 2 table* like the one below, which applies to our data.

	Exp.	Control	Σ's
Above	a 14	b 6	20
Below	c 6	d 14	20
Σ's	20 (+)	20 (=)	40 = n

This looks very much like the 2 × 2 table used for calculating χ^2 in a test of independence (p. 192). It is, and it suggests the next step.

Step 3: Calculate χ^2. This may be done by Method II of Table 26 (p. 193); but it is better to *use formula 15.4 with the correction for discontinuity* (p. 198). This is easy to use because all of the marginal sums in the denominator will have the same value. For example, with our data:

$$\chi_c^2 = \frac{n(|ad - bc| - n/2)^2}{(a + b)(c + d)(a + c)(b + d)} \qquad (15.4)$$

$$= \frac{40(|14 \times 14 - 6 \times 6| - 40/2)^2}{20 \times 20 \times 20 \times 20} = 4.90$$

Step 4: Interpret χ^2 with the help of the χ^2 table (p. 188). In this case we use $d.f. = 1$. Our *null hypothesis* in the median test is that the samples (two in this case) come from the same, or identical, populations. If this were true, we should have the same number of scores above (and below) the grand median in both groups, except for chance variations in sampling. In using χ^2 to test this hypothesis we are in effect substituting the hypothesis that the (relative) number of scores above and below the grand median is *independent of the groups involved*, experimental or control. From Table 25, for $d.f. = 1$, we find that our obtained value of χ^2, 4.90 > 3.841, the tabled value for $p = .05$. We may, therefore, reject H_o at the $p < .05$ level and conclude that the proportion of positive and negative deviations from the median is *not independent of*, but rather is

TABLE 31

A Median Test for Four Treatment Groups Using a χ^2 Test of Independence in a 2 × 4 Table

	Group 1	Group 2	Group 3	Group 4	Σ's
Above	1a 12 (10)	2a 6 (10)	3a 7 (10)	4a 15 (10)	40
Below	1b 8 (10)	2b 14 (10)	3b 13 (10)	4b 5 (10)	40
Σ's	20	20	20	20	80 = n

A. *Determination of Theoretical Frequencies* (f_t)

Cell	Calculation	f_t	Cell	Calculation	f_t
1a	$\dfrac{40 \times 20}{80}$	10	1b	$\dfrac{40 \times 20}{80}$	10

We can now see that f_t *must be the same (10) for all cells*, because with the same number of scores in each group the products of the marginal totals will be the same for all cells.

B. *Calculation of* χ^2

χ^2 is the sum of the $(f_o - f_t)^2/f_t$ ratios for all cells. In this case

$$\chi^2 = \frac{(12-10)^2 + (6-10)^2 + 3^2 + 5^2 + 2^2 + 4^2 + 3^2 + 5^2}{10} = 10.8$$

C. *Interpretation*

Using $d.f. = (r-1)(c-1) = 1 \times 3 = 3$ in the χ^2 table (p. 188), we find that the tabled value of χ^2 is 9.837 for $p = .02$. Since $10.8 > 9.837$, we may reject H_o at the $p < .02$ level. For the nature of H_o and further implications of the test see text.

associated with, the groups. In terms of the original null hypothesis, we may conclude that the experimental and control group samples come from different populations.

As in the case of the H test, *the median test may be applied to more than two groups as a rough kind of preliminary analysis of variance*. To do this we may make use of the χ^2 test of independence again. But in this case we use a $2 \times k$ table similar to Table 27 (p. 194). We shall illustrate this with four experimental groups of 20 cases each. The *grand median* has already been calculated and the distribution of positive and negative deviations from this is given in Table 31, along with the χ^2 calculations.

Implications of the Median Test Used with More than Two Groups.

The *null hypothesis* in the four-group illustration in Table 31 is the same as that for two groups: namely, that the four groups are random samples from the same, or identical, populations. In using the χ^2 test of independence as a test of this hypothesis we introduced a somewhat more specific hypothesis: the relative number of scores above and below the grand median is independent of the treatment groups. Since the probability of the obtained χ^2 in our illustration in Table 31 was so small, we rejected this hypothesis. We may conclude, therefore, that the patterns of positive and negative deviations from the grand median are *not independent of, but are associated with, the treatment groups* in some way. Or we may conclude that *not all the groups come from the same population*. But if we are using the median test as a kind of *simple analysis of variance*, the question arises: *Which groups do not conform*? Or, *which pair of groups* shows such different patterns as to make it seem improbable that they have a common origin? Unfortunately, there is no immediate answer to these questions. There is no Scheffé test to help us out. The best we can do is to inspect the 2×4 table, pick out the two groups whose patterns of positive and negative deviations are most markedly different (e. g., groups 2 and 4), and then get the complete set of scores for these two groups and run a new significance test on these two groups alone. If the result is significant, we can try again with two groups that are a little less different; and so on and on. A fairly dreary procedure! This absence of something like a Scheffé test to pinpoint significant comparisons between two groups when the preliminary test tells us that differences exist *somewhere* in the k groups is one of the weaknesses of the median test. Other comparative weaknesses will be mentioned in the next section.

4. The Median Test Compared with Other Significance Tests

The median test, like the H rank test, is sometimes appropriate when there is reason to suspect that the parent populations are far from normally distributed. It is also easy to carry out. But we have just seen one of its weaknesses, which it shares with the H test. If it is used with more than two groups as a kind of preliminary analysis of variance, there is *no equivalent of the Scheffé test* for locating the source of probable differences from the common population. And *neither of these order tests can be used in two-factor analysis*, which is possible with the F test.

A basic weakness in the median test is the *vague character of the null hypotheses* which must be used with it. This results in equally *vague conclusions* if H_o is rejected; e.g., "The populations from which the samples come are not identical," or "The patterns of positive and negative deviations from the grand median (whatever they are) are not independent of the groups." Whereas, with the t test based on exact scores, we may reach more specific conclusions, such as $\mu_1 > \mu_2$. Or we may calculate power for a specific risk of a Type I error and a specific $\mu_1 - \mu_2$. This vagueness in the conclusions of the median test stems from the fact that it makes only minimal use of the information contained in the original data, even less use than is made by the rank test; whereas the t and the F tests, based on exact scores, make much more complete use of the data.

For the same reason the median test has *less power* than the t and F tests, even less than the H test. For example, in our two-group illustration of the median test above the best we could do was to reject H_o at the $p < .05$ level; whereas calculations show that a t test making more complete use of the same data would make it possible to reject H_o at the $p < .01$ level. In other words, in using the less sensitive median test we would have been more likely to overlook a true experimental effect. If the differences had been just a little smaller, we might well have made a Type II error.

There are *two other minor drawbacks* to the median test. (1) The χ^2 distribution gives *probabilities* which *only approximate* the true probabilities, especially for small theoretical frequencies. It should not be used unless n is at least 20 and unless the theoretical frequencies are at least 5 for each cell. The observed frequencies may be smaller. (2) The median test *assumes that tied scores will not occur*. If they occur *at the grand median*, they will effect the χ^2 values. The safest procedure is to allocate such ties in such a way that the difference between the positive and negative frequencies will be as small as possible.

Final Comment: *We are never forced by the nature of the data to use the median test in preference to the t or F tests*, because in all these cases exact scores must be available to start with. The median test may be used, however, for its limited purposes, if our samples suggest that the parent populations are far from normally distributed; or it may be used for quick checking purposes. But *in general* it is better to trust to the robust character of the t and F tests. (For character reference, see p. 82.)

Exercises

1. Apply the Kruskal-Wallis H test of significance to the sets of scores below from three different experimental treatments, using the indicated steps (over):

	(1)	(2)	(3)
	1	2	7
	0	5	6
	4	3	9
	2	6	10
	3	4	8
($\bar{X}$):	(2.0)	(4.0)	(8.0)

(a) Rank all 15 scores and get the sum of the ranks for each of the three groups. (b) Apply the accuracy check to these three sums. (c) Find the sum of the T_k^2/n_k values. (d) Find H. (e) Interpret H, giving H_o and conclusion.

2. Apply the H test to the 10 scores in sets (2) and (3) above, using the steps corresponding to steps (a) to (e) in Exercise 1. (f) Is the difference between the means ($\bar{X}_3 - \bar{X}_2$) probably significant?

3. Apply the H test to the 10 scores in sets (1) and (3) of Exercise 1, using the steps corresponding to steps (a) to (e). (f) Is $\bar{X}_3 - \bar{X}_1$ probably significant?

4. Using the H test with sets (1) and (2) of Exercise 1, find out if $\bar{X}_2 - \bar{X}_1$ is significant. Indicate all steps.

5. The scores in sets (1) and (2) of Exercise 1 are combined into a single group of 10 which is to be compared with the five scores in set (3). Apply the H test, showing all steps, and draw an appropriate conclusion about the means of the two groups.

6. Using the H test, see if the 20 scores below from a mechanical ingenuity test differ significantly for the random samples of college men and college women. Show all the usual steps, (a) to (e) of Exercise 1. (f) Which sample probably comes from a superior population with respect to mechanical ingenuity? What about other kinds of ingenuity?

Men: 3, 5, 6, 9, 7, 6, 8, 10, 9, 7 ($\bar{X} = 7.0$)
Women: 6, 3, 5, 4, 2, 0, 2, 4, 1, 3 ($\bar{X} = 3.0$)

7. In an experimental group 11 scores lie above and 5 below the grand median. In a control group 11 scores lie below and 5 above this same point. (a) Calculate χ_c^2 for the median test. (b) State an appropriate H_o. (c) What conclusion do you draw based on the tabled value of χ^2 for $p = .05$ and the proper d.f.?

8. Given the following distribution of positive and negative deviations from the grand median for four treatment groups. Apply the median test using the indicated steps: (a) Calculate χ^2. (b) State an appropriate H_o. (c) Draw an appropriate conclusion.

Groups:	1	2	3	4
Positive:	6	5	13	16
Negative:	14	15	7	4

9. The original scores for Groups 1 and 3 in Exercise 8 are as follows. Apply the median test to these two groups. What conclusion do you draw?

Group 1: 10 9 8 7 7 6 6 6 6 6 5 5 5 4 4 4 3 3 2
Group 3: 11 10 10 9 9 9 8 8 8 8 7 7 7 7 6 6 5 4 3

TABLE 32

Table of Squares and Square Roots of Numbers From 1 to 1000

Number	Square	Square Root	Number	Square	Square Root
1	1	1.000	31	9 61	5.568
2	4	1.414	32	10 24	5.657
3	9	1.732	33	10 89	5.745
4	16	2.000	34	11 56	5.831
5	25	2.236	35	12 25	5.916
6	36	2.449	36	12 96	6.000
7	49	2.646	37	13 69	6.083
8	64	2.828	38	14 44	6.164
9	81	3.000	39	15 21	6.245
10	1 00	3.162	40	16 00	6.325
11	1 21	3.317	41	16 81	6.403
12	1 44	3.464	42	17 64	6.481
13	1 69	3.606	43	18 49	6.557
14	1 96	3.742	44	19 36	6.633
15	2 25	3.873	45	20 25	6.708
16	2 56	4.000	46	21 16	6.782
17	2 89	4.123	47	22 09	6.856
18	3 24	4.243	48	23 04	6.928
19	3 61	4.359	49	24 01	7.000
20	4 00	4.472	50	25 00	7.071
21	4 41	4.583	51	26 01	7.141
22	4 84	4.690	52	27 04	7.211
23	5 29	4.796	53	28 09	7.280
24	5 76	4.899	54	29 16	7.348
25	6 25	5.000	55	30 25	7.416
26	6 76	5.099	56	31 36	7.483
27	7 29	5.196	57	32 49	7.550
28	7 84	5.292	58	33 64	7.616
29	8 41	5.385	59	34 81	7.681
30	9 00	5.477	60	36 00	7.746

Table of Squares and Square Roots (Continued)

Number	Square	Square Root	Number	Square	Square Root
61	37 21	7.810	91	82 81	9.539
62	38 44	7.874	92	84 64	9.592
63	39 69	7.937	93	86 49	9.644
64	40 96	8.000	94	88 36	9.695
65	42 25	8.062	95	90 25	9.747
66	43 56	8.124	96	92 16	9.798
67	44 89	8.185	97	94 09	9.849
68	46 24	8.246	98	96 04	9.899
69	47 61	8.307	99	98 01	9.950
70	49 00	8.367	100	1 00 00	10.000
71	50 41	8.426	101	1 02 01	10.050
72	51 84	8.485	102	1 04 04	10.100
73	53 29	8.544	103	1 06 09	10.149
74	54 76	8.602	104	1 08 16	10.198
75	56 25	8.660	105	1 10 25	10.247
76	57 76	8.718	106	1 12 36	10.296
77	59 29	8.775	107	1 14 49	10.344
78	60 84	8.832	108	1 16 64	10.392
79	62 41	8.888	109	1 18 81	10.440
80	64 00	8.944	110	1 21 00	10.488
81	65 61	9.000	111	1 23 21	10.536
82	67 24	9.055	112	1 25 44	10.583
83	68 89	9.110	113	1 27 69	10.630
84	70 56	9.165	114	1 29 96	10.677
85	72 25	9.220	115	1 32 25	10.724
86	73 96	9.274	116	1 34 56	10.770
87	75 69	9.327	117	1 36 89	10.817
88	77 44	9.381	118	1 39 24	10.863
89	79 21	9.434	119	1 41 61	10.909
90	81 00	9.487	120	1 44 00	10.954

Squares and Square Roots

Table of Squares and Square Roots (Continued)

Number	Square	Square Root	Number	Square	Square Root
121	1 46 41	11.000	151	2 28 01	12.288
122	1 48 84	11.045	152	2 31 04	12.329
123	1 51 29	11.091	153	2 34 09	12.369
124	1 53 76	11.136	154	2 37 16	12.410
125	1 56 25	11.180	155	2 40 25	12.450
126	1 58 76	11.225	156	2 43 36	12.490
127	1 61 29	11.269	157	2 46 49	12.530
128	1 63 84	11.314	158	2 49 64	12.570
129	1 66 41	11.358	159	2 52 81	12.610
130	1 69 00	11.402	160	2 56 00	12.649
131	1 71 61	11.446	161	2 59 21	12.689
132	1 74 24	11.489	162	2 62 44	12.728
133	1 76 89	11.533	163	2 65 69	12.767
134	1 79 56	11.576	164	2 68 96	12.806
135	1 82 25	11.619	165	2 72 25	12.845
136	1 84 96	11.662	166	2 75 56	12.884
137	1 87 69	11.705	167	2 78 89	12.923
138	1 90 44	11.747	168	2 82 24	12.961
139	1 93 21	11.790	169	2 85 61	13.000
140	1 96 00	11.832	170	2 89 00	13.038
141	1 98 81	11.874	171	2 92 41	13.077
142	2 01 64	11.916	172	2 95 84	13.115
143	2 04 49	11.958	173	2 99 29	13.153
144	2 07 36	12.000	174	3 02 76	13.191
145	2 10 25	12.042	175	3 06 25	13.229
146	2 13 16	12.083	176	3 09 76	13.266
147	2 16 09	12.124	177	3 13 29	13.304
148	2 19 04	12.166	178	3 16 84	13.342
149	2 22 01	12.207	179	3 20 41	13.379
150	2 25 00	12.247	180	3 24 00	13.416

Table of Squares and Square Roots (Continued)

Number	Square	Square Root	Number	Square	Square Root
181	3 27 61	13.454	211	4 45 21	14.526
182	3 31 24	13.491	212	4 49 44	14.560
183	3 34 89	13.528	213	4 53 69	14.595
184	3 38 56	13.565	214	4 57 96	14.629
185	3 42 25	13.601	215	4 62 25	14.663
186	3 45 96	13.638	216	4 66 56	14.697
187	3 49 69	13.675	217	4 70 89	14.731
188	3 53 44	13.711	218	4 75 24	14.765
189	3 57 21	13.748	219	4 79 61	14.799
190	3 61 00	13.784	220	4 84 00	14.832
191	3 64 81	13.820	221	4 88 41	14.866
192	3 68 64	13.856	222	4 92 84	14.900
193	3 72 49	13.892	223	4 97 29	14.933
194	3 76 36	13.928	224	5 01 76	14.967
195	3 80 25	13.964	225	5 06 25	15.000
196	3 84 16	14.000	226	5 10 76	15.033
197	3 88 09	14.036	227	5 15 29	15.067
198	3 92 04	14.071	228	5 19 84	15.100
199	3 96 01	14.107	229	5 24 41	15.133
200	4 00 00	14.142	230	5 29 00	15.166
201	4 04 01	14.177	231	5 33 61	15.199
202	4 08 04	14.213	232	5 38 24	15.232
203	4 12 09	14.248	233	5 42 89	15.264
204	4 16 16	14.283	234	5 47 56	15.297
205	4 20 25	14.318	235	5 52 25	15.330
206	4 24 36	14.353	236	5 56 96	15.362
207	4 28 49	14.387	237	5 61 69	15.395
208	4 32 64	14.422	238	5 66 44	15.427
209	4 36 81	14.457	239	5 71 21	15.460
210	4 41 00	14.491	240	5 76 00	15.492

Squares and Square Roots

Table of Squares and Square Roots (Continued)

Number	Square	Square Root	Number	Square	Square Root
241	5 80 81	15.524	271	7 34 41	16.462
242	5 85 64	15.556	272	7 39 84	16.492
243	5 90 49	15.588	273	7 45 29	16.523
244	5 95 36	15.620	274	7 50 76	16.553
245	6 00 25	15.652	275	7 56 25	16.583
246	6 05 16	15.684	276	7 61 76	16.613
247	6 10 09	15.716	277	7 67 29	16.643
248	6 15 04	15.748	278	7 72 84	16.673
249	6 20 01	15.780	279	7 78 41	16.703
250	6 25 00	15.811	280	7 84 00	16.733
251	6 30 01	15.843	281	7 89 61	16.763
252	6 35 04	15.875	282	7 95 24	16.793
253	6 40 09	15.906	283	8 00 89	16.823
254	6 45 16	15.937	284	8 06 56	16.852
255	6 50 25	15.969	285	8 12 25	16.882
256	6 55 36	16.000	286	8 17 96	16.912
257	6 60 49	16.031	287	8 23 69	16.941
258	6 65 64	16.062	288	8 29 44	16.971
259	6 70 81	16.093	289	8 35 21	17.000
260	6 76 00	16.125	290	8 41 00	17.029
261	6 81 21	16.155	291	8 46 81	17.059
262	6 86 44	16.186	292	8 52 64	17.088
263	6 91 69	16.217	293	8 58 49	17.117
264	6 96 96	16.248	294	8 64 36	17.146
265	7 02 25	16.279	295	8 70 25	17.176
266	7 07 56	16.310	296	8 76 16	17.205
267	7 12 89	16.340	297	8 82 09	17.234
268	7 18 24	16.371	298	8 88 04	17.263
269	7 23 61	16.401	299	8 94 01	17.292
270	7 29 00	16.432	300	9 00 00	17.321

Table of Squares and Square Roots (Continued)

Number	Square	Square Root	Number	Square	Square Root
301	9 06 01	17.349	331	10 95 61	18.193
302	9 12 04	17.378	332	11 02 24	18.221
303	9 18 09	17.407	333	11 08 89	18.248
304	9 24 16	17.436	334	11 15 56	18.276
305	9 30 25	17.464	335	11 22 25	18.303
306	9 36 36	17.493	336	11 28 96	18.330
307	9 42 49	17.521	337	11 35 69	18.358
308	9 48 64	17.550	338	11 42 44	18.385
309	9 54 81	17.578	339	11 49 21	18.412
310	9 61 00	17.607	340	11 56 00	18.439
311	9 67 21	17.635	341	11 62 81	18.466
312	9 73 44	17.664	342	11 69 64	18.493
313	9 79 69	17.692	343	11 76 49	18.520
314	9 85 96	17.720	344	11 83 36	18.547
315	9 92 25	17.748	345	11 90 25	18.574
316	9 98 56	17.776	346	11 97 16	18.601
317	10 04 89	17.804	347	12 04 09	18.628
318	10 11 24	17.833	348	12 11 04	18.655
319	10 17 61	17.861	349	12 18 01	18.682
320	10 24 00	17.889	350	12 25 00	18.708
321	10 30 41	17.916	351	12 32 01	18.735
322	10 36 84	17.944	352	12 39 04	18.762
323	10 43 29	17.972	353	12 46 09	18.788
324	10 49 76	18.000	354	12 53 16	18.815
325	10 56 25	18.028	355	12 06 25	18.841
326	10 62 76	18.055	356	12 67 36	18.868
327	10 69 29	18.083	357	12 74 49	18.894
328	10 75 84	18.111	358	12 81 64	18.921
329	10 82 41	18.138	359	12 88 81	18.947
330	10 89 00	18.166	360	12 96 00	18.974

Squares and Square Roots

Table of Squares and Square Roots (Continued)

Number	Square	Square Root	Number	Square	Square Root
361	13 03 21	19.000	391	15 28 81	19.774
362	13 10 44	19.026	392	15 36 64	19.799
363	13 17 69	19.053	393	15 44 49	19.824
364	13 24 96	19.079	394	15 52 36	19.849
365	13 32 25	19.105	395	15 60 25	19.875
366	13 39 56	19.131	396	15 68 16	19.900
367	13 46 89	19.157	397	15 76 09	19.925
368	13 54 24	19.183	398	15 84 04	19.950
369	13 61 61	19.209	399	15 92 01	19.975
370	13 69 00	19.235	400	16 00 00	20.000
371	13 76 41	19.261	401	16 08 01	20.025
372	13 83 84	19.287	402	16 16 04	20.050
373	13 91 29	19.313	403	16 24 09	20.075
374	13 98 76	19.339	404	16 32 16	20.100
375	14 06 25	19.363	405	16 40 25	20.125
376	14 13 76	19.391	406	16 48 36	20.149
377	14 21 29	19.416	407	16 56 49	20.174
378	14 28 84	19.442	408	16 64 64	20.199
379	14 36 41	19.468	409	16 72 81	20.224
380	14 44 00	19.494	410	16 81 00	20.248
381	14 51 61	19.519	411	16 89 21	20.273
382	14 59 24	19.545	412	16 97 44	20.298
383	14 66 89	19.570	413	17 05 69	20.322
384	14 74 56	19.596	414	17 13 96	20.347
385	14 82 25	19.621	415	17 22 25	20.372
386	14 89 96	19.647	416	17 30 56	20.396
387	14 97 69	19.672	417	17 38 89	20.421
388	15 05 44	19.698	418	17 47 24	20.445
389	15 13 21	19.723	419	17 55 61	20.469
390	15 21 00	19.748	420	17 64 00	20.494

Table of Squares and Square Roots (Continued)

Number	Square	Square Root	Number	Square	Square Root
421	17 72 41	20.518	451	20 34 01	21.237
422	17 80 84	20.543	452	20 43 04	21.260
423	17 89 29	20.567	453	20 52 09	21.284
424	17 97 76	20.591	454	20 61 16	21.307
425	18 06 25	20.616	455	20 70 25	21.331
426	18 14 76	20.640	456	20 79 36	21.354
427	18 23 29	20.664	457	20 88 49	21.378
428	18 31 84	20.688	458	20 97 64	21.401
429	18 40 41	20.712	459	21 06 81	21.424
430	18 49 00	20.736	460	21 16 00	21.448
431	18 57 61	20.761	461	21 25 21	21.471
432	18 66 24	20.785	462	21 34 44	21.494
433	18 74 89	20.809	463	21 43 69	21.517
434	18 83 56	20.833	464	21 52 96	21.541
435	18 92 25	20.857	465	21 62 25	21.564
436	19 00 96	20.881	466	21 71 56	21.587
437	19 09 69	20.905	467	21 80 89	21.610
438	19 18 44	20.928	468	21 90 24	21.633
439	19 27 21	20.952	469	21 99 61	21.656
440	19 36 00	20.976	470	22 09 00	21.679
441	19 44 81	21.000	471	22 18 41	21.703
442	19 53 64	21.024	472	22 27 84	21.726
443	19 62 49	21.048	473	22 37 29	21.749
444	19 71 36	21.071	474	22 46 76	21.772
445	19 80 25	21.095	475	22 56 25	21.794
446	19 89 16	21.119	476	22 65 76	21.817
447	19 98 09	21.142	477	22 75 29	21.840
448	20 07 04	21.166	478	22 84 84	21.863
449	20 16 01	21.190	479	22 94 41	21.886
450	20 25 00	21.213	480	23 04 00	21.909

Squares and Square Roots

Table of Squares and Square Roots (Continued)

Number	Square	Square Root	Number	Square	Square Root
481	23 13 61	21.932	511	26 11 21	22.605
482	23 23 24	21.954	512	26 21 44	22.627
483	23 32 89	21.977	513	26 31 69	22.650
484	23 42 56	22.000	514	26 41 96	22.672
485	23 52 25	22.023	515	26 52 25	22.694
486	23 61 96	22.045	516	26 62 56	22.716
487	23 71 69	22.068	517	26 72 89	22.738
488	23 81 44	22.091	518	26 83 24	22.760
489	23 91 21	22.113	519	26 93 61	22.782
490	24 01 00	22.136	520	27 04 00	22.804
491	24 10 81	22.159	521	27 14 41	22.825
492	24 20 64	22.181	522	27 24 84	22.847
493	24 30 49	22.204	523	27 35 29	22.869
494	24 40 36	22.226	524	27 45 76	22.891
495	24 50 25	22.249	525	27 56 25	22.913
496	24 60 16	22.271	526	27 66 76	22.935
497	24 70 09	22.293	527	27 77 29	22.956
498	24 80 04	22.316	528	27 87 84	22.978
499	24 90 03	22.338	529	27 98 41	23.000
500	25 00 00	22.361	530	28 09 00	23.022
501	25 10 01	22.383	531	28 19 61	23.043
502	25 20 04	22.405	532	28 30 24	23.065
503	25 30 09	22.428	533	28 40 89	23.087
504	25 40 16	22.450	534	28 51 56	23.108
505	25 50 25	22.472	535	28 62 25	23.130
506	25 60 36	22.494	536	28 72 96	23.152
507	25 70 49	22.517	537	28 83 69	23.173
508	25 80 64	22.539	538	28 94 44	23.195
509	25 90 81	22.561	539	29 05 21	23.216
510	26 01 00	22.583	540	29 16 00	23.238

Squares and Square Roots

Table of Squares and Square Roots (Continued)

Number	Square	Square Root	Number	Square	Square Root
541	29 26 81	23.259	571	32 60 41	23.896
542	29 37 64	23.281	572	32 71 84	23.917
543	29 48 49	23.302	573	32 83 29	23.937
544	29 59 36	23.324	574	32 94 76	23.958
545	29 70 25	23.345	575	33 06 25	23.979
546	29 81 16	23.367	576	33 17 76	24.000
547	29 92 09	23.388	577	33 29 29	24.021
548	30 03 04	23.409	578	33 40 84	24.042
549	30 14 01	23.431	579	33 52 41	24.062
550	30 25 00	23.452	580	33 64 00	24.083
551	30 36 01	23.473	581	33 75 61	24.104
552	30 47 04	23.495	582	33 87 24	24.125
553	30 58 09	23.516	583	33 98 89	24.145
554	30 69 16	23.537	584	34 10 56	24.166
555	30 80 25	23.558	585	34 22 25	24.187
556	30 91 36	23.580	586	34 33 96	24.207
557	31 02 49	23.601	587	34 45 69	24.228
558	31 13 64	23.622	588	34 57 44	24.249
559	31 24 81	23.643	589	34 69 21	24.269
560	31 36 00	23.664	590	34 81 00	24.290
561	31 47 21	23.685	591	34 92 81	24.310
562	31 58 44	23.707	592	35 04 64	24.331
563	31 69 69	23.728	593	35 16 49	24.352
564	31 80 96	23.749	594	35 28 36	24.372
565	31 92 25	23.770	595	35 40 25	24.393
566	32 03 56	23.791	596	35 52 16	24.413
567	32 14 89	23.812	597	35 64 09	24.434
568	32 26 24	23.833	598	35 76 04	24.454
569	32 37 61	23.854	599	35 88 01	24.474
570	32 49 00	23.875	600	36 00 00	24.495

Squares and Square Roots

Table of Squares and Square Roots (Continued)

Number	Square	Square Root	Number	Square	Square Root
601	36 12 01	24.515	631	39 81 61	25.120
602	36 24 04	24.536	632	39 94 24	25.140
603	36 36 09	24.556	633	40 06 89	25.159
604	36 48 16	24.576	634	40 19 56	25.179
605	36 60 25	24.597	635	40 32 25	25.199
606	36 72 36	24.617	636	40 44 96	25.219
607	36 84 49	24.637	637	40 57 69	25.239
608	36 96 64	24.658	638	40 70 44	25.259
609	37 08 81	24.678	639	40 83 21	25.278
610	37 21 00	24.698	640	40 96 00	25.298
611	37 33 21	24.718	641	41 08 81	25.318
612	37 45 44	24.739	642	41 21 64	25.338
613	37 57 69	24.759	643	41 34 49	25.357
614	37 69 96	24.779	644	41 47 36	25.377
615	37 82 25	24.799	645	41 60 25	25.397
616	37 94 56	24.819	646	41 73 16	25.417
617	38 06 89	24.839	647	41 86 09	25.436
618	38 19 24	24.860	648	41 99 04	25.456
619	38 31 61	24.880	649	42 12 01	25.475
620	38 44 00	24.900	650	42 25 00	25.495
621	38 56 41	24.920	651	42 38 01	25.515
622	38 68 84	24.940	652	42 51 04	25.534
623	38 81 29	24.960	653	42 64 09	25.554
624	38 93 76	24.980	654	42 77 16	25.573
625	39 06 25	25.000	655	42 90 25	25.593
626	39 18 76	25.020	656	43 03 36	25.612
627	39 31 29	25.040	657	43 16 49	25.632
628	39 43 84	25.060	658	43 29 64	25.652
629	39 56 41	25.080	659	43 42 81	25.671
630	39 69 00	25.100	660	43 56 00	25.690

Table of Squares and Square Roots (Continued)

Number	Square	Square Root	Number	Square	Square Root
661	43 69 21	25.710	691	47 74 81	26.287
662	43 82 44	25.729	692	47 88 64	26.306
663	43 95 69	25.749	693	48 02 49	26.325
664	44 08 96	25.768	694	48 16 36	26.344
665	44 22 25	25.788	695	48 30 25	26.363
666	44 35 56	25.807	696	48 44 16	26.382
667	44 48 89	25.826	697	48 58 09	26.401
668	44 62 24	25.846	698	48 72 04	26.420
669	44 75 61	25.865	699	48 86 01	26.439
670	44 89 00	25.884	700	49 00 00	26.458
671	45 02 41	25.904	701	49 14 01	26.476
672	45 15 84	25.923	702	49 28 04	26.495
673	45 29 29	25.942	703	40 42 09	26.514
674	45 42 76	25.962	704	49 56 16	26.533
675	45 56 25	25.981	705	49 70 25	26.552
676	45 69 76	26.000	706	49 84 36	26.571
677	45 83 29	26.019	707	49 98 49	26.589
678	45 96 84	26.038	708	50 12 64	26.608
679	46 10 41	26.058	709	50 26 81	26.627
680	46 24 00	26.077	710	50 41 00	26.646
681	46 37 61	26.096	711	50 55 21	26.665
682	46 51 24	26.115	712	50 69 44	26.683
683	46 64 89	26.134	713	50 83 69	26.702
684	46 78 56	26.153	714	50 97 96	26.721
685	46 92 25	26.173	715	51 12 25	26.739
686	47 05 96	26.192	716	51 26 56	26.758
687	47 19 69	26.211	717	51 40 89	26.777
688	47 33 44	26.230	718	51 55 24	26.796
689	47 47 21	26.249	719	51 69 61	26.814
690	47 61 00	26.268	720	51 84 00	26.833

Squares and Square Roots

Table of Squares and Square Roots (Continued)

Number	Square	Square Root	Number	Square	Square Root
721	51 98 41	26.851	751	56 40 01	27.404
722	52 12 84	26.870	752	56 55 04	27.423
723	52 27 29	26.889	753	56 70 09	27.441
724	52 41 76	26.907	754	56 85 16	27.459
725	52 56 25	26.926	755	57 00 25	27.477
726	52 70 76	26.944	756	57 15 36	27.495
727	52 85 29	26.963	757	57 30 49	27.514
728	52 99 84	26.981	758	57 45 64	27.532
729	53 14 41	27.000	759	57 60 81	27.550
730	53 29 00	27.019	760	57 76 00	27.568
731	53 43 61	27.037	761	57 91 21	27.586
732	53 48 24	27.055	762	58 06 44	27.604
533	53 72 89	27.074	763	58 21 69	27.622
734	53 87 56	27.092	764	58 36 96	27.641
735	54 02 25	27.111	765	58 52 25	27.659
736	54 16 96	27.129	766	58 67 56	27.677
737	54 31 69	27.148	767	58 82 89	27.695
738	54 46 44	27.166	768	58 98 24	27.713
739	54 61 21	27.185	769	59 13 61	27.731
740	54 76 00	27.203	770	59 29 00	27.749
741	54 90 81	27.221	771	59 44 41	27.767
742	55 05 64	27.240	772	59 59 84	27.785
743	55 20 49	27.258	773	59 75 29	27.803
744	55 35 36	27.276	774	59 90 76	27.821
745	55 50 25	27.295	775	60 06 25	27.839
746	55 65 16	27.313	776	60 21 76	27.857
747	55 80 09	27.331	777	60 37 29	27.875
748	55 95 04	27.350	778	60 52 84	27.893
749	56 10 01	27.368	779	60 68 41	27.911
750	56 25 00	27.386	780	60 84 00	27.928

Squares and Square Roots

Table of Squares and Square Roots (Continued)

Number	Square	Square Root	Number	Square	Square Root
781	60 99 61	27.946	811	65 77 21	28.478
782	61 15 24	27.964	812	65 93 44	28.496
783	61 30 89	27.982	813	66 09 69	28.513
784	61 46 56	28.000	814	66 25 96	28.531
785	61 62 25	28.018	815	66 42 25	28.548
786	61 77 96	28.036	816	66 58 56	28.566
787	61 93 69	28.054	817	66 74 89	28.583
788	62 09 44	28.071	818	66 91 24	28.601
789	62 25 21	28.089	819	67 07 61	28.618
790	62 41 00	28.107	820	67 24 00	28.636
791	62 56 81	28.125	821	67 40 41	28.653
792	62 72 64	28.142	822	67 56 84	28.671
793	62 88 49	28.160	823	67 73 29	28.688
794	63 04 36	28.178	824	67 89 76	28.705
795	63 20 25	28.196	825	68 06 25	28.723
796	63 36 16	28.213	826	68 22 76	28.740
797	63 52 09	28.231	827	68 39 29	28.758
798	63 68 04	28.249	828	68 55 84	28.775
799	63 84 01	28.267	829	68 72 41	28.792
800	64 00 00	28.284	830	68 89 00	28.810
801	64 16 01	28.302	831	69 05 61	28.827
802	64 32 04	28 320	832	69 22 24	28.844
803	64 48 09	28.337	833	69 38 89	28.862
804	64 64 16	28.355	834	69 55 56	28.879
805	64 80 25	28.373	835	69 72 25	28.896
806	64 96 36	28.390	836	69 88 96	28.914
807	65 12 49	28.408	837	70 05 69	28.931
808	65 28 64	28.425	838	70 22 44	28.948
809	65 44 81	28.443	839	70 39 21	28.965
810	65 61 00	28.460	840	70 56 00	28.983

Table of Squares and Square Roots (Continued)

Number	Square	Square Root	Number	Square	Square Root
841	70 72 81	29.000	871	75 86 41	29.513
842	70 89 64	29.017	872	76 03 84	29.530
843	71 06 49	29.034	873	76 21 29	29.547
844	71 23 36	29.052	874	76 38 76	29.563
845	71 40 25	29.069	875	76 56 25	29.580
846	71 57 16	29.086	876	76 73 76	29.597
847	71 74 09	29.103	877	76 91 29	29.614
848	71 91 04	29.120	878	77 08 84	29.631
849	72 08 01	29.138	879	77 26 41	29.648
850	72 25 00	29.155	880	77 44 00	29.665
851	72 42 01	29.172	881	77 61 61	29.682
852	72 59 04	29.189	882	77 79 24	29.698
853	72 76 09	29.206	883	77 96 89	29.715
854	72 93 16	29.223	884	78 14 56	29.732
855	73 10 25	29.240	885	78 32 25	29.749
856	73 27 36	29.257	886	78 49 96	29.766
857	73 44 49	29.275	887	78 67 69	29.783
858	73 61 64	29.292	888	78 85 44	29.799
859	73 78 81	29.309	889	79 03 21	29.816
860	73 96 00	29.326	890	79 21 00	29.833
861	74 13 21	29.343	891	79 38 81	29.850
862	74 30 44	29.360	892	79 56 64	29.866
863	74 47 69	29.377	893	79 74 49	29.883
864	74 64 96	29.394	894	79 92 36	29.900
865	74 82 25	29.411	895	80 10 25	29.916
866	74 99 56	29.428	896	80 28 16	29.933
867	75 16 89	29.445	897	80 46 09	29.950
868	75 34 24	29.462	898	80 64 04	29.967
869	75 51 61	29.479	899	80 82 01	29.983
870	75 69 00	29.496	900	81 00 00	30.000

Squares and Square Roots 231

Table of Squares and Square Roots (Continued)

Number	Square	Square Root	Number	Square	Square Root
901	81 18 01	30.017	931	86 67 61	30.512
902	81 36 04	30.033	932	86 86 24	30.529
903	81 54 09	30.050	933	87 04 89	30.545
904	81 72 16	30.067	934	87 23 56	30.561
905	81 90 25	30.083	935	87 42 25	30.578
906	82 08 36	30.100	936	87 60 96	30.594
907	82 26 49	30.116	937	87 79 69	30.610
908	82 44 64	30.133	938	87 98 44	30.627
909	82 62 81	30.150	939	88 17 21	30.643
910	82 81 00	30.166	940	88 36 00	30.659
911	82 99 21	30.183	941	88 54 81	30.676
912	83 17 44	30.199	942	88 73 64	30.692
913	83 35 69	30.216	943	88 92 49	30.708
914	83 53 96	30.232	944	89 11 36	30.725
915	83 72 25	30.249	945	89 30 25	30.741
916	83 90 56	30.265	946	89 49 16	30.757
917	84 08 89	30.282	947	89 68 09	30.773
918	84 27 24	30.299	948	89 87 04	30.790
919	84 45 61	30.315	949	90 06 01	30.806
920	84 64 00	30.332	950	90 25 00	30.822
921	84 82 41	30.348	951	90 44 01	30.838
922	85 00 84	30.364	952	90 63 04	30.854
923	85 19 29	30.381	953	90 82 09	30.871
924	85 37 76	30.397	954	91 01 16	30.887
925	85 56 25	30.414	955	91 20 25	30.903
926	85 74 76	30.430	956	91 39 36	30.919
927	85 93 29	30.447	957	91 58 49	30.935
928	86 11 84	30.463	958	91 77 64	30.952
929	86 30 41	30.480	959	91 96 81	30.968
930	86 49 00	30.496	960	92 16 00	30.984

Squares and Square Roots

Table of Squares and Square Roots (Concluded)

Number	Square	Square Root	Number	Square	Square Root
961	92 35 21	31.000	981	96 23 61	31.321
962	92 54 44	31.016	982	96 43 24	31.337
963	92 73 69	31.032	983	96 62 89	31.353
964	92 92 96	31.048	984	96 82 56	31.369
965	93 12 25	31.064	985	97 02 25	31.385
966	93 31 56	31.081	986	97 21 96	31.401
967	93 50 89	31.097	987	97 41 69	31.417
968	93 70 24	31.113	988	97 61 44	31.432
969	93 89 61	31.129	989	97 81 21	31.448
970	94 09 00	31.145	990	98 01 00	31.464
971	94 28 41	31.161	991	98 20 81	31.480
972	94 47 84	31.177	992	98 40 64	31.496
973	94 67 29	31.193	993	98 60 49	31.512
974	94 86 76	31.209	994	98 80 36	31.528
975	95 06 25	31.225	995	99 00 25	31.544
976	95 25 76	31.241	996	99 20 16	31.559
977	95 45 29	31.257	997	99 40 09	31.575
978	95 64 84	31.273	998	99 60 04	31.591
979	95 84 41	31.289	999	99 80 01	31.607
980	96 04 00	31.305	1000	100 00 00	31.623

Selected Answers to Exercises

Chapter 2

1. Frequencies from bottom step (30–34) to top step (70–74): 3 3 7 11 9 7 4 3 1

4. Yes. No, but a little skewed.

5. Step designations as in Table 1. Frequencies from bottom up: 1 2 1 2 7 8 5 6 5 3 4 1 2 1

8. No; too many steps used. Curve pretty erratic.

9. Frequencies from step 30–34 to step 70–74: 1 2 4 7 2 4 2 1 1. (Other step intervals possible.)

11. Frequencies from step 30–34 to step 65–69: 2 1 3 4 7 3 2 2. (Other step intervals possible.)

13. Opposite. No. Small samples.

Chapter 3

1. (a) 50.1 (b) 50.1

3. (a) 49.0 (b) 49.5
5. 52
7. $(42 + 46)/2 = 44$; 46
9. $3(-12)/42 + 52 = 51.14$
11. 53.05

Chapter 4

1. 34
3. 6.3
6. 88.2, 9.4 (for $n = 46$)
7. Based on $c = 52$ in Exercise 6: $199 = 159 - 6 + 46 = 199$
9. $MD = 7.8$ (for $n = 10$)
11. 91.3, 9.55 (for $n = 10$)
13. 52.2, 7.23 (for $n = 40$)

Chapter 5

1. 38.1, 51.7, 64.5
3. 37, 47, 57, 67
5. 40.9, 45.1, 58.0
7. 37, 47, 57, 67
9. 36, 44, 52, 60, 68

Chapter 6

1. 1.32
3. Arithmetic $T = 24$, reading $T = 32$
5. 30, 70, 60, 45, 40, 65, 55, 50
7.
Subject:	1	2	3	4	5	6	7	8
Mean T score:	36.5	63.5	58.5	59	45	46	54	48.5
Rank:	8	1	3	2	7	6	4	5

   Order is different because T score method makes more complete use of data than rank method; hence, gives more accurate results.

Chapter 7

1. 50–70, 58–74, 50–90
2. (a) 79.6 (b) 85.8 (c) 40.4–79.6

3. (a) 76.24 (b) 52.84 (c) 60.6–71.4
5. (a) 15.9% (b) 84.1% (c) 10% (d) 90%
7. (a) .200 (b) .100 (c) .020 (d) .001
9. (a) 66.08–73.92 (b) 64.84 − 75.16 (c) 66.71 − 73.29

Chapter 8

1. (a) 3.0 (b) 2.67 (c) $\mu_1 = \mu_2$ (d) $p < .02$ (e) Reject H_o (at the $p < .02$ level) and conclude that $\mu_2 > \mu_1$ (girls' population probably superior).

3. (a) 2.67 (b) 3.00 (c) $\mu_1 = \mu_2$ (d) $p < .01$ (e) Reject H_o (at the $p < .01$ level) and conclude $\mu_2 > \mu_1$ (f) Makes a significant difference more likely.

5. (a) 1.30 (b) 3.86 (c) $\mu_1 = \mu_2$ (d) $p < .01$ (e) Reject H_o (at the $p < .01$ level) and conclude boys' population probably superior to girls'. This modifies the conclusion in Exercise 4 in which H_o was retained.

7. 2.94. No.

Chapter 9

1. (a) 2.0 (b) 1.8 (c) $\mu_1 = \mu_2$ (d) $p > .05$ (e) Retain H_o and conclude means do not differ significantly. Two tails because we are interested in a difference in either direction.

3. (a) 2.0 (b) 2.35 (c) $\mu_1 = \mu_2$ (d) $p < .02$ (e) Reject H_o (at the $p < .02$ level) and conclude that means differ significantly. Two-tailed test because direction of difference not specified.

5. ($\Sigma x_1^2 = 39{,}600$, $\Sigma x_2^2 = 17{,}700$) (a) 3.0 (b) 3.33 (c) $\mu_1 = \mu_2$ (d) $p < .002$ (e) Reject H_o (at the $p < .002$ level) and conclude that means differ significantly. Two-tailed.

7. 9900 (a) 2.00 (b) 2.4 (c) $\mu_1 = \mu_2$ (d) $p < .02$ (e) Difference is significant (at the $p < .02$ level).

Chapter 10

1, 3. See text.

5. (1) 3.92, − 3.92 (2) 0, − 3.92 (3) .50, .00 (4) .50, .50

7. (a) 1.00 (b) .40 (c) .60
9. 168

Chapter 11

1. (a) 200 (b) 120 (c) 80 (d) 27, 2 (e) 4.45, 40, 9.0 (f) $\mu_1 = \mu_2 = \mu_3$. Since $9.0 > 5.49$ reject H_o and conclude that there is a highly significant overall experimental effect (at the $p < .01$ level).

3. (a) .94 (b) 2.59, 3.32 (c) 2.44, 3.12 (d) $\bar{X}_3 - \bar{X}_1$ (at the $p < .01$ level).

5. (a) $(1)\bar{X}_3 + (-\frac{1}{2})\bar{X}_1 + (-\frac{1}{2})\bar{X}_2$ and $1 - \frac{1}{2} - \frac{1}{2} = 0$ (b) .816 (c) 3.68 (d) 2.59, 3.32 (e) Since $3.68 > 3.32$, $\bar{X}_3$ is significantly greater than the average of $\bar{X}_1$ and $\bar{X}_2$ (at the $p < .01$ level).

7. (a) $(1)\bar{X}_3 + (-\frac{1}{3})\bar{X}_1 + (-\frac{1}{3})\bar{X}_2 + (-\frac{1}{3})\bar{X}_4$ and $1 - \frac{1}{3} - \frac{1}{3} - \frac{1}{3} = 0$ (b) .913 (c) 4.38 (d) 3.12, 3.98 (e) Since $4.38 > 3.98$, $\bar{X}_3$ is significantly greater (at the $p < .01$ level.).

Chapter 12

1. (a) 5,30, 25, 20; 1.0, 6.0, 5.0, 4.0; 25, 190, 135, 90 (b) 120, 70, 50 (c) 7.45. $\mu_{11} = \mu_{12} = \mu_{21} = \mu_{22}$. Since $7.45 > 5.29$ there is a significant overall experimental effect at the $p < .01$ level.

3. (a) 1.12 (b) 3.12, 3.98 (c) 3.49, 4.46 (d) $\bar{X}_{12} - \bar{X}_{11}$ at the $p < .01$ level and $\bar{X}_{21} - \bar{X}_{11}$ at the $p < .05$ level.

5. (a) 20, 67.5, and (since from Ex. 4 $SS_B = 95$) 7.5 (b) (Using $MS_W = 2.0$ from Ex. 4) 10.0, 11.25, 1.25 (c) Since $10.0 > 7.00$ (by interpolation), there is a significant main effect of A at the $p < .01$ level. Since $11.25 > 4.07$ (by interpolation), there is a significant main effect of B at the $p < .01$ level. Since $1.25 < 2.74$, the interaction effect is not significant.

Chapter 13

1. $-.22$
3. $-.17$

5. .97
7. .92
9. Obtained value of t much less than 2.12, the tabled value of t for $d.f. = 16$ and $p = .05$. It is also less than 1.96, the value of z for $p = .05$. Hence, retain H_o that X,Y pairs are random samples from population with zero correlation between X and Y, and conclude that correlation is not significantly different from zero.

Chapter 14

1. .79 ($A = 12{,}682$, $B = 4776$, $C = 54{,}424$)
3. (a) 24, 27; 54, 65 (b) 44 (c) 75×18.1, 25×31.0, 225×35.8; .54
5. (a) 81.8, 158.2 (b) 0.17, 1.75 (c) $\tilde{X} = 0.17Y + 54.9$, $\tilde{Y} = 1.75X + 15.0$
7. (a) 81.8, 157.6 (b) 0.233, 2.66 (c) $\tilde{X} = 0.233Y + 45.1$, $\tilde{Y} = 2.66X - 60.0$
9. (a) 80 (b) 4.47 (c) 68.5 and 91.5
10. (a) $-.10$ (b) No

Chapter 15

1. 4.33. Since this is less (much less) than the tabled value of χ^2 for $d.f. = 4$ and $p = .05$ (9.488), we retain the hypothesis that the grades are normally distributed.
3. (a) 5 (b) 17, 10
5. (a,b)

	Athletes	Nonathletes	Σ's
Good Adjustment	a 20 (15)	b 30 (35)	50
Fair Adjustment	c 10 (15)	d 40 (35)	50
Σ's	30	70	100

(c) $1.67 + 0.71 + 1.67 + 0.71 = 4.76$ (d) 4.76 (e) $4.76 > 3.841$ (for $d.f. = 1$ and $p = .05$). Hence, reject hypothesis and conclude that adjustment is not independent of (but associated with) athletic classification.

7. (a) Row 1: 7, 21, 43, 21, 8 Row 2: 21, 63, 129, 63, 24 Row 3: 7, 21, 43, 21, 8 (b) 37.69 (c) 37.69 > 20.09, the tabled value for $d.f. = 8$ and $p = .01$. Hence, reject hypothesis and conclude that grades are not independent of (but associated with) adjustment.

Chapter 16

1. (a) 60.5, 44.0, 15.5 (b) 120 = 120 (c) 1167.75 (d) 10.39 (e) Since $10.39 > 9.21$, reject H_o (at the $p < .01$ level) and conclude samples come from different populations.

3. (a) 40, 15 (b) 55 = 55 (c) 365 (d) 6.8 (e) Since $6.8 > 6.635$, we reject H_o (at the $p < .01$ level) and conclude the two samples come from different populations. (f) Yes.

5. (a) 104.5, 15.5 (b) 120 = 120 (c) 1140.08 (d) 9.0 (e) Since $9.0 > 6.635$, we reject H_o (at the $p < .01$ level) and conclude samples come from different populations. (f) $\bar{X}_3$ significantly greater than the mean of the combined groups.

7. (a) 3.13 (b) Populations are identical (c) Retain H_o.

9. Grand median: 6.5. Patterns: 5+ and 15−; 15+ and 5−. χ_c^2 $8.1 > 7.815$ for $p = .05$. Therefore, conclude that populations differ.

Index

A

Analysis of variance, 113-153
  simple analysis, 113-132
    calculation of F, 124
    calculation of SS's, 121-124
    interpretation of F, 120-121, 124-125
    main steps, 114-115
    Scheffé comparisons, 125-131
  two-factor analysis, 133-153
    calculation of F, 137, 142, 148
    calculation of SS's, 135-137, 139-142
    components of SS_B, 139-142, 146-148
    first stage, 135-138
    fixed effects and other experimental designs, 151-152
    interaction, 133-134, 138, 140, 142-144, 147, 149-151
    interpretation, 138, 143-144, 147
    Scheffé comparisons, 149-151
    second stage, 138-151
Association, test of, 191-197

B

Bias, correction for, 39
Binomial distribution, 15n

C

Centile (*see* Percentile)
Central tendency, measures of,
Central tendency, (*continued*) 19-32
  the crude mode, 22
  the mean, 20-22, 26-30
    see also Mean
  the median, 22-26
Chance, curve of, 15
Charlier's check, 44
Combining test results, 56-62
Comparing test results, 56-62
Comparisons of means, in analysis of variance, 113-114, 121, 125-131, 144
  complex comparisons in Scheffé test, 128-131
  standard error of, 130-131
Chi-square distribution (*see* Chi-square test)
  and Kruskal-Wallis H test, 204-206
  and median test, 209-211, 213
  and phi correlation, 183
Chi-square test, 185-199
  calculation of χ^2 statistic, 186-187, 192-193
  correction for discontinuity, 197-198
  degrees of freedom in, 189-191, 197-198
  and goodness of fit, 186-189
  interpretation of χ^2, 187-189
  significance test for phi coefficient, 183
  and tests of independence (or association), 191-197
  use of χ^2 table, 187-189

Coded scores, calculation of mean from, 26–30
  calculation of Σx^2 from, 40–43
  calculation of standard deviation and variance from, 40–43
  in correlation chart, 168–172
Confidence, in significance tests, 86, 125
Confidence limits, for test scores, 162–163
  for population mean, 70–73
Control group, 4–5, 145
Correlation techniques, 154–184
  calculation by correlation chart, 167–172
  and confidence limits for test scores, 162–163
  interpretation of coefficients, 159–160
  machine calculation, 165–167
  phi coefficient of correlation, 180–183
  predictions from regression equations, 173–180
  product-moment method, 157–158, 165–172
  rank-difference method, 155–157
  and reliability of tests, 160–161
  significance test for r, 157–159
  and validity of tests, 161–162
Crude mode, 22

D

Data, grouped and ungrouped, 10, 20
Decile norms, 52
Decision making, 98–111
  risks in, 99
  and test power, 99, 102–106
  and Type I and Type II errors, 99, 102–106, 108–110
Degrees of freedom
  in analysis of variance, 115, 125, 137, 139, 142, 147–148
  in chi-square tests, 189–191, 197–198
  in H test, 203–204
  in Kruskal-Wallis rank test, 203–204

Degrees of freedom (*continued*)
  in median test, 209–210
  in t test, 79
Derivation of formulas, for mean from coded scores, 31–32
  for Σfx^2 for coded scores, 48–49
  for Σx^2 for uncoded scores, 47–48
Descriptive statistics, 38

E

Edwards, A. L., see Preface
Equating groups, 4
Estimate, of population mean, 70–73
  of population variance, 76, 110, 120, 126
  of position of random scores, 69–70
  of sample size for specified test power, 107–110
  standard error of, 159–160, 176, 179–181
Experimental design, 110–111, 145, 151–152
Experimental effects, 121, 125, 138, 147, 151–152
Experimentation, in physics and psychology, 2–3
  and testing, 6–7

F

F distributions, 115–120
  table of F, 116–119
Fisher, R. A., 81, 83, 187–188
Fixed effects experimental design, 151–152
Forecasting efficiency, index of, 180–181
F ratio, calculation of, 124, 137, 142, 148
  general nature of, 115–120
  interpretation of in analysis of variance, 120–121, 124–125, 138, 143–144, 147
  table of F, 116–119
Frequency (*see* Frequency distributions)
  polygon, 12–13
  table, 9–12

Frequency distributions
  bimodal, 16
  binomial distribution, 15n
  see Chi-square distribution
  see F distributions
  graphs of, 12–14
  see Normal distribution curve
  skewed, 16
  see t distribution
  see z distribution

G

Goodness of fit, 186–189
Grading "on the curve," 53–54
Grouped data, 10–11

H

Hays, W. L., see Preface, 152n
Histogram, 12–14
H test of significance, 202–207
Hypothesis testing (see Chi-square test; see Null hypothesis)

I

Independence, test of, 191–197
Individual differences, 34
Inferential statistics, 39
Interaction in analysis of variance, 138, 140, 142–144, 147

J

J distribution, 14n

K

Kruskal-Wallis rank test of significance, 202–207, 212
  calculation of H, 204–205
  comparison with F and t tests, 206–207
  critique of, 202–203
  interpretation of H, 204–206
  use as simple analysis of variance, 202
Kruskal, W. H., 202–203

L

Limits of confidence (see Confidence limits)

M

Machine calculation, derivation of formula for Σx^2, 47–48
  of the mean, 20
  of product-moment correlation, 165–167
  of regression coefficients, 167, 177–178
  of Σx^2, 39–40
  of the standard deviation, 39–40
  of SS's in variance analysis, 142
  of the variance, 39–40
Mean, 20–22, 26–32
  calculation from coded data, 26–30
  calculation from grouped data, 20–22
  calculation from ungrouped data, 20
  confidence limits for, 70–73
  derivation of coding formulas, 30–32
  of population and sample, 70–72
  standard error of, 71–72
  variance of, 71
  versus median, 24–26
Mean deviation, 35–37, 46
  calculation of, 37
  relation to Q and s, 46
Mean squares in analysis of variance, 115–119, 121, 124–126, 130, 137, 142, 144, 147–148, 150
Median, 22–24
  calculation from grouped data, 23–24
  calculation from ungrouped data, 23
  versus mean, 24–26
Median test of significance, 207–213
  calculations required, 208–210
  comparison with other significance tests, 212–213
  interpretation, 209–212
  use with more than two groups, 211–212

Mixed effects experimental design, 151–152
Mode, crude, 22

N

Nonparametric tests, 201
 see Order significance tests
Normal distribution curve, 14–17
 see z distribution curve
Normal probability curve, 14–17
 see z distribution curve
Norms, 51–54
Null hypothesis, 78–79, 81–82, 86
 in analysis of variance, 120–121, 124–125
 in calculating test power, 99–100, 102–105
 in correlated samples, 86
 criteria for rejecting, 81–82
 in estimating sample size, 107–108
 in increasing test power, 106–108
 in Kruskal-Wallis rank test, 204–207
 in median test, 209–213
 in one-tailed and two-tailed tests, 92–94
 in significance of correlations, 157, 183

O

One-tailed significance test
 meaning of, 92–93
 power of, 94–95
 when to use, 92, 94
Order of merit rankings, 66–67
Order significance tests, 201–215
 see Kruskal-Wallis rank test
 see Median test

P

Parameter, 39, 72
Parametric tests, 201
Pearson, K., 157, 186
Percentile, norms, 52
 rank equivalents, 61–62
Population of scores, 39
Power of significance tests, 94–95, 98–110
 calculating power, 102–106

Power of significance tests (continued)
 increasing power, 106–110
 of one- and two-tailed tests, 94–95
 relative power of order and classical significance tests, 207, 212
Prediction, and confidence limits, 69–73, 162–163
 and degrees of confidence, 69, 72
 and forecasting efficiency, 180
 from normal curve, 69
 from regression equations, 173–180
 from standard error of measurement, 162–163
Probability,
 and chi-square tests, 187–189, 193–195
 and confidence limits, 70–73
 estimates, 65, 69–73, 75
 and F distribution, 120–121, 124–125, 127, 131, 138, 144, 147
 and order tests, 204–207, 210
 and quartile deviation, 46
 and regression equations, 176–179
 and significance tests, 64–65
 and statistics, 2
 and t distribution, 75, 78–82
 and test power, 98–110
 and Type I and Type II errors, 81–82, 99–110
 and variability, 2, 46
 and z distribution, 64–74, 89–90, 100–110

Q

Quartile deviation, 44–46
 calculation of, 45
 and probability, 46
 relation to MD, and s, 46
Quartile norms, 52

R

Random effects experimental design, 151–152
Random samples, in analysis of variance, 115–121
 in chi-square test, 185, 187
 in confidence limits, 70–72

Random samples (*continued*)
  in random effects design, 152
  in *t* test, 78–81
Range, 35
Rank-difference correlation, 155–157
Rank test of significance, 202–207
Regression, coefficients, 174–175, 177–178
  equations, 173–180
  line, 174, 177
Rejection criteria in significance tests, 81–82, 91–92, 94
  see Decision making
Reliability of tests, 160–161
Risks in significance tests, 99
Robust test, 82

S

Samples, desirability of large, 5-6
  and population, 39
  see Random samples
  size and test power, 107–110
Scatter diagram, 168–169, 179
Scheffé, H., 125–130, 144, 149–150, 202, 212
Scheffé test of significance, common standard error of difference, 126–127, 130, 144, 150
  for complex comparisons, 128–131
  criterion of significance, 126–128, 131
  for difference between two means, 125–127, 144, 149–151
  minimum difference for significance, 128, 144, 150
  standard error of comparison, 130–131
Science, statistical nature of, 1–2
Scores, definition of, 8–9
  distribution of, 8–17
  relative nature of, 51–52, 57
  standard, 56–61
  T scores, 59–61
Seitz, C. P., 85
Sigma scores (*see* Standard scores)
Significance tests, analysis of variance, 113–114, 120–121, 124–128, 138, 143–144, 147
  chi-square tests, 185–198

Significance tests (*continued*)
  for difference between means, 75–86, 89–90
  H test, 202–207
  Kruskal-Wallis rank test, 202–207
  median test, 207–213
  one-tailed and two-tailed, 90–95, 120
  for phi coefficient of correlation, 183
  power of, 94–95, 98–110, 207, 212
  for product-moment correlation, 157–159
  Scheffé test, 125–131, 144, 149–151
  statistical significance vs. importance, 110–111
  the t test, 75–87
  the z test, 89–90, 100–102
Smith, G. M., 85
Spearman-Brown formula for test reliability, 161
Standard deviation, 36–44
  calculation from coded data, 40–43
  calculation from correlation chart, 169–171
  calculation by machine, 39–40
  calculation from ungrouped data, 37–38
  and probability, 46–47
  relation to Q and MD, 46–47
  and standard error of estimate, 176, 179–180
Standard error, of a comparison, 130–131
  of a difference between means, 77–78, 82–83, 90, 101–105
  of estimate, 159–160, 176, 179–181
  of the mean, 70–71
  of the mean difference, 84–85
  of measurement, 162–163
  of an obtained score, 162–163
  in Scheffé test, 125–127, 130, 144, 150
Standard normal curve (*see* z distribution curve)
Standard scores, 56–61
  converted to T scores, 59–61
  normalized, 60–61

Statistic, a, 39
Stelazine, 145–151
Step interval, 10–11
Summation, rules of, 30
Sum of squares, in analysis of variance, 121–124, 135–137, 139–142
 in correlation calculations, 158, 170
 definition, 38
 derivation of formulas for, 47–49
 in regression coefficients, 175
 in standard deviation formula, 38
 in variance formula, 38

T

Tally of scores, 9–10
t distribution, 79–81
 relation to z distribution, 90
 table of t, 80
 use in decision making and demining test power, 100
 use in significance tests, 75–87, 89–95
Testing and experimentation, 6–7
Thorazine, 145–151
Treatment effects, 147
Treatment groups, 114–115, 120–121
T scores, 59–61
t test, 75–87
 and analysis of variance, 113–114
 calculation of t, 77–78, 82–86
 for correlated samples, 83–87
 for correlation coefficients, 157–159
 evaluation of t, 78–82, 86
 for large samples, 89–91
 one-tailed and two-tailed tests, 90–95
 in Scheffé test, 125–128, 131, 144, 150
 for small samples, 76–83
Two-tailed significance test, meaning of, 91
 power of, 94–95
 when to use, 91–92
Type I and Type II errors, 81–82, 94–95
 and decision making, 94–95, 99, 102–106

Type I and Type II errors (*continued*)
 and one-tailed and two-tailed tests, 94–95
 and test power, 94–95, 99–100, 102–110

U

U distribution, 14n

V

Validity of tests, 161–162
Variable, continuous, 8–9, 180
 dichotomous, 181
Variability, measures of, 33–49
 and probability, 2
Variance, 36–44
 analysis of, 113–153
 calculation from coded scores, 40–44
 calculation by machine, 39–40
 calculation from ungrouped scores, 37–38
 and F ratio, 115–121
 of the mean, 70–71
 see Mean squares
 and test power, 100, 106, 110

W

Wallis, W. A., 202–203

Y

Yates, F., 197

Z

z distribution curve, areas under, 65–70
 and calculating test power, 100–110
 and confidence limits, 70–73
 predictions from, 69–70
 and probability, 64–74, 89–90, 100–110
 relation to t distribution, 90
 and significance tests, 89–90, 100–102
 table of z, 68
z scores (*see* Standard scores)
z test, 89–90, 100–102